Halphen Distribution Family

WITH APPLICATION IN HYDROLOGICAL FREQUENCY ANALYSIS

Salaheddine El Adlouni and Bernard Bobée

Since 1971
Water Resources Publications, LLC WRP Colorado

For Information and Correspondence:

Water Resources Publications, LLC
P. O. Box 630026, Highlands Ranch, Colorado 80163-0026, USA

HALPHEN DISTRIBUTION FAMILY
WITH APPLICATION IN HYDROLOGICAL FREQUENCY ANALYSIS

Salaheddine El Adlouni
Professor
Université de Moncton, Department of Mathematic & Statistics,
Moncton, NB E1A 3E9, Canada

and

Bernard Bobée
Emeritus Professor
Institut National de la Recherche Scientifique (INRS-ETE)
Ville de Québec, QC G1K 9A9, Canada

ISBN 13: 978-1887201902
Library of Congress Control Number: 2016958765

Cover Designed by WRP

This publication is printed and bound in the United States of America.

Foreword

Efficient management and adequate estimation of natural resources are important human preoccupations in 2016. What the authors call "Frequency analysis" and we prefer to call "probabilistic analysis" is an essential basis for a quantified representation of natural phenomena such as river flows, rainfall, etc. The range of available probability models for this purpose is very broad. The authors propose a set of such models which is not new but which, in 2016, answers to an ever present need. Let's start with a little part of the history of the twentieth century in France. Founded in 1912, the SHF (Société Hydrotechnique de France) was an original creation grouping together the functions of a classical scientific society and, with its Technical Committee, those of a technical studies service available to producers and distributors of electricity, which were private up to the nationalization of 1946.

From 1939 and during the first years of the world war, it was headed by Pierre Massé, future Électricité de France (EDF) president and Commissioner of Planning (1965-1969). Eminent economist and creator of Dynamic Programming before Bellman, Massé was convinced of the importance of probabilistic and statistical techniques for the management of hydroelectric resources. In 1940 he created a Hydrology Statistics Service inside SHF gathering a number of outstanding researchers. One of them was Etienne Halphen. During this war period, the SHF laboratories were in free French "zone" and therefore Massé and SHF could protect Halphen against the "hunt for Jews." Then in 1946, the SHF teams formed the nucleus of the Department of Studies and Hydraulic Research of EDF and the LNH (Laboratoire National d'Hydraulique). Halphen worked there until his untimely death in 1954, both on statistical theory and applications in resource estimation, planning and management of hydropower, and the risks associated with floods. In 1947 and 1948, he pushed EDF to engage two young researchers who are not unknown to the statistical world: namely Lucien Le Cam and Georges Morlat. In 1950 Le Cam left EDF for the USA, encouraged to come to Berkeley University by Jerzy Neyman.

Etienne Halphen (1911-1954), former student from the Ecole Normale Supérieure, was a classmate and friend of Daniel Dugué. *An anecdote: a discourse on the laws of Halphen was presented by Dugué at the Academy of Sciences in the early 1940s, but after the Liberation of France, Dugué restored the authorship of this work to Halphen.* Actually the first results on the Harmonic distribution appeared in 1941, on the laws of HA in 1946, and in 1048, on the law HB.

The first writings were from the 1950's and the laws of HIB were added to the system in 1956, by an engineer from the team; M Larcher. These three classes then constituted a "complete system" as the system of the Pearson laws. In his time, Halphen had already largely access to the range of classical distributions, but he was convinced that none of these forms of distribution was generally adequate for all series of hydrometric observations from French rivers (for any time, day, month, year, with the minimum and maximum, etc.). Indeed the theoretical and applied works on "Halphen laws" were continued until the end of 1960. However, since 1960 very few published works were using this very interesting and important family of distributions. Mathematically, these distributions are defined by 3 parameters, the minimum considered as essential by Halphen to represent the range of hydrological regimes. In competition with the Pearson laws or transformed normal laws (such as Cox-Box), both having important theoretical and practical advantages as introduced in the present work.

The difficulty of implementation was for a long time the lack of tabulations. We must greatly thank MM. El Adlouni and Bobée for their very beautiful, insightful, and clear presentation of Halphen laws. Furthermore, they provide the first tools for their practical implementation through the use of modern Monte Carlo algorithms. With these tools, the analytical forms of law A, B, IB are no longer barriers to the simulation, generation of samples, and estimation of Halphen distributions. Being particularly concerned with the applications to maxima, the authors while paying attention to the asymptotic properties of these distributions, the end result is that their behavior is very helpful, as compared with other distributions.

In matters of inference, if the theory of maximum likelihood is presented, only the method of moments is applied and incorporated into their "Decision Support System" and HYFRAN-PLUS software. The first is a system with both graphical and computational tools, facilitating looking for the best fit to given samples. The generation of samples from Halphen laws as entered into the software is made according to the acceptance-rejection algorithm. The corresponding Matlab programs are added in annexes. The philosophy of general statistical presentation is essentially classical. Indeed the frequentist interpretation of probability is favoured in this monograph as the title attests. However, the Theory of Probability propose another practical interpretation that is the bayesian subjective one, where probability is considered as a "personal degree of rational belief", given to the occurrence of an event or the validity of a claim. It is this interpretation to be adopted when the authors present the bayesian estimation of the parameters of Halphen laws in Chapter 5. Recent developments on MCMC algorithms give us hope to see the application of these bayesian estimation and inference methods.

We hope that this presentation, at the same time clear and theoretically solid, and offering modern tools for the application of these distribution models, will allow a wide dissemination of these models among hydrologists and other specialists in Earth Sciences.

August, 2016
Jacques Bernier,
St-Martial de Nabirat

Preface

Statistical analysis is increasingly used in water resources management and planning. Frequency analysis is a particularly useful tool to estimate the probability of occurrence of a given event by fitting a statistical distribution to a sample of independent and identically distributed data.

For estimation of extreme events, such as floods, we are facing two problems: the small size of the sample of observed data and the choice of the most adequate distribution to represent the tail behavior of the variable.

When we are working on the content of the software HYFRAN, Professor Jacques Bernier, a regular visiting scientist of the Chair of Statistical Hydrology (CHS), suggested to us to consider using the Halphen family of distributions. Despite their inferential properties (sufficiency of the estimators and the existence of a complete system) these distributions were, at this time, relatively less well known and hardly used, due to the complexity of the computational aspects related to normalization constants.

This text brings together theoretical works made in last fifteen years and includes Matlab codes for Halphen fitting and generating.

Chapter 1 introduces the subject. In Chapters 2 to 4, each of the three Halphen distributions, are presented and discussed successively:

- mathematical and statistical properties;
- asymptotic behavior and model selection;
- parameter estimation methods.

Chapter 5 and 6 are concerned with the simulation and generation of samples from Halphen distributions and the Matlab codes. While our readership is expected large in the hydrological community of scientists and engineers, basic statistical level is required.

We take pleasure in thanking firstly, Professor Jacques Bernier for inspiriting this research and for his support and comments for an earlier version of this effort. Without his help and suggestions, this work would never have been possible. We would also like to thank him for writing the foreword to this book. In addition, we are also grateful to Rémy Garçon for sending us references concerning Halphen distributions and to our colleagues: Taha Ouarda and Pierre Hubert for the stimulating discussions we had on the use of Halphen distributions for hydrological studies.

Many thanks to colleagues, assistants and research associates, co-authors of several publications; especially Luc Perreault and Fateh Chebana.

We would like to express our gratitude to Ms. Branka McLaughlin for her editorial work, our contact with her has always been pleasant and constructive.

January, 2017
S.E.A., Moncton
B.B., Québec

Table of Contents

List of Tables

List of Figures

Chapter 1
Introduction to Frequency Analysis Using Halphen Distributions

In many fields (such as hydrology, meteorology, finance, internet traffic, forest fire, wind speed, etc.), the occurrence of extreme events cannot be forecasted on the basis of deterministic information with sufficient skill and lead-time. In such cases, a probabilistic approach is required to incorporate the effects of such phenomena into decisions. For example, in hydrology, extreme events such as floods, droughts, and storms have significant economic consequences and can cause loss of human life. Frequency Analysis (FA) is of substantial importance for reservoir management or dam design. A key element in the FA is the estimation of the probability of exceedance $p = \Pr(X \geq x_T)$ of the event x_T corresponding to the return period T. The event x_T of return period T corresponds to the quantile of probability of exceedance $p = 1/T = \Pr(X \geq x_T)$, thus for a given probability distribution F,

$$\Pr(X \leq x_T) = F(x_T;\underline{\theta}) = 1 - 1/T \tag{1.1}$$

and

$$x_T = F^{-1}(1 - 1/T;\underline{\theta}) \tag{1.2}$$

where $\underline{\theta}$ is the vector of parameters. Note that p corresponds generally to the risk and T is the mean *(over a large period, theoretically infinite)* of the time units separating two successive events.

Several distributions are commonly used in FA and are generally divided into three groups: the normal family (normal, Lognormal); the General Extreme Value (GEV) family (Gumbel, Fréchet, and reverse Weibull); and the Pearson type 3 family (Gamma, Pearson type 3, Log-Pearson type 3). However, the use of the GEV distribution is currently a standard, thanks to its explicit quantile function $x_T(F)$ and the availability of software for parameter estimation. The GEV distribution appears to be a universal model for extremes in many fields. In hydrology, several countries of the world (for example, the United

Kingdom) consider the GEV for floods as standard model (Robson and Reed, 1999). The same kind of standardization is recommended using the Log-Pearson type 3 distribution in the USA and Australia or Lognormal distribution in China (Bobée, 1999). However, "*it is one thing to standardize manufacturing techniques and another to attempt this with natural phenomena*" (Alexander et al. 1970).

Several other distributions were developed and applied to estimate extreme events, but their use remains very limited. Halphen (1941) introduced the Halphen type A (HA) distribution to fit a large variety of data sets and to obtain additional flexibility when compared to the lognormal distribution (also called Gibrat-Gauss). Indeed, in this era, the normality of the logarithmic transformation was a common assumption. Gibrat (1931) proposed a mathematical justification for the use of lognormal distribution based on proportional effect law. In some cases, the logarithmic transformation is enough to get a symmetric distribution (the normal case). However, some empirical studies show that the symmetry of the logarithmic transformation is not satisfactory. Even in cases where the symmetry of the logarithm is verified, it would appear, for many cases, that the density of the lognormal distribution do not decrease fast enough, especially for large values. To solve this problem, Halphen (1941) proposed HA distribution which has a lighter tail than the lognormal distribution. Later, Halphen (1955) defined type B (HB) distribution to obtain greater flexibility for modeling the smaller values in a data set. Halphen's research was motivated by the following two reasons: **(F)** flexibility and availability of several probability distribution shapes, especially for the right tail (extreme values); and **(S)** sufficiency of parameter estimators (optimal estimation with minimum variance). These two reasons will be detailed in Chapters 3 and 4. Morlat (1956) reported that Larcher (researcher at Electricité de France) proposed the Halphen Inverse B distribution (HIB) to complete the Halphen system of distributions and presented some of its statistical and mathematical properties. Note that the Gamma and Inverse Gamma distributions are limiting cases of the Halphen distributions which will be discussed later in the text.

All the above groups of distributions are deduced from mathematical links between the distributions of each group and less attention has been given to their asymptotic behaviors. Quantile estimation for small return periods is almost the same for all distributions commonly used in hydrological frequency analysis; however, significant differences are observed for large return periods (larger than the double of the sample size), mainly due to the uncertainties involved in extrapolation. El

Adlouni et al. (2008) present a classification of the distributions, generally used in FA, with respect to their tail behavior. Almost all distributions given in this classification, to fit the annual maximum series, belong to the Class D of sub-exponential distributions or the Class C, with regularly varying tail (Embrechts et al., 2003). More theoretical detail concerning these classes are given in Chapter 3. Halphen and GEV systems cover both classes. Indeed, the HA, HB, Gamma, and Gumbel distributions belong to the Class D (Sub-exponential distributions) while HIB, Inverse Gamma and Fréchet, belong to the Class C (regularly varying distributions). El Adlouni and Bobée (2010) presented some graphical tools to discriminate between these classes, which are available in the Decision Support System of the HYFRAN-PLUS software:
(http://www.wrpllc.com/books/hyfran.html).

This book aims to gather the research studies concerning the Halphen distributions family; starting from the first research done by Halphen (1941), those of Morlat (1956), and then all the recent developments from the group of the chair in statistical hydrological (CHS, 2002).

Although this book is essentially dedicated to the Halphen family, other distributions, frequently used in Hydrological Frequency Analysis (HYFRAN-PLUS), have been mentioned for a better understanding.

The following chapters will give detail information concerning the Halphen distribution family. Chapter 2 will present the mathematical and statistical properties of the three Halphen distributions. In Chapter 3, we will look at the asymptotic behavior of the Halphen distributions and compare them to the models commonly used in hydrological frequency analysis. Chapter 4 deals with the parameter estimation problems. Chapter 5 presents the algorithms for generating random samples from the Halphen distributions. The last chapter, Chapter 6, is devoted to the presentation of the Matlab codes for sample generation from Halphen distributions, parameter estimation, and quantile computation. More detail on the functions of Bessel and exponential factorial are presented in Appendices A and B, respectively. Appendix C describes the Decision Support System (DSS) to select the distribution that fits the data the best, especially for extremes. In addition these appendices present the properties of the dispersion functions that are important for the resolution of maximum likelihood systems.

Adlouni et al. (2008) present a classification of the distributions, generally used in FA, with respect to their tail behavior. Almost all distributions given in this classification to fit the annual maximum series, belong to one (Class D of sub-exponential distributions), the Class C [illegible] (Chebana et al., 2007) [illegible] [illegible] Chapter 3, the generalized [illegible]

[illegible] distribution [illegible] exponential distributions [illegible] HB, Inverse [illegible] and [illegible] Class [illegible] distributions [illegible] Adl[illegible] [illegible] which [illegible] [illegible] tail behavior [illegible] of the [illegible] suffices [illegible] commonly used [illegible].

[illegible] aims to gather the research studies concerning the Halphen distributions family starting from the first research done by Halphen (1941), those of [illegible] (1955) and the more recent developments from the group of the chair in statistical hydrology [illegible] (C[illegible] 2002).

Although this book is essentially dedicated to the Halphen family, other distributions frequently used in hydrological Frequency Analysis (HYFRAN-PLUS) have been considered for a better understanding.

The following chapters will give detail information concerning the Halphen distribution family. Chapter 2 will present the mathematical and statistical properties of the three Halphen distributions. In Chapter 3, we will look at the asymptotic behavior of the Halphen distributions and compare them to other models commonly used in hydrological frequency analysis. Chapter [illegible] deals with the parameter estimation problems. Chapter 5 presents the algorithms for generating random samples from the Halphen distributions. The last chapter, Chapter 6, is devoted to the presentation of the Matlab codes for sample generation from Halphen distributions, parameter estimation, and quantile computation. More details on the functions of Bessel and exponential factorial are presented in Appendices A and B, respectively. Appendix C describes the Decision Support System (DSS) to select the distribution that fits the data the best, especially for extremes. In addition, these appendices present the properties of the dispersion functions that are important for the resolution of maximum likelihood systems.

Chapter 2
Halphen Distribution Family: Mathematical and Statistical Properties

2.1. HISTORICAL CONTEXT

The Halphen family of distributions has been specifically designed to model annual and seasonal flows (Morlat, 1956) and to fit a large variety of data set. The first investigations of Halphen were motivated by the Plan program for global hydropower resources. So the problem of Halphen was related to the estimation of the annual average and seasonal flow of the rivers in France (Le Cam and Morlat, 1949; Labaye, 1956). The first use of the Halphen distribution for extremes was considered as comparative study of the Halphen type A distribution with the Extreme Value distributions (Bernier, 1959). These distributions are characterized by:

- A lower bound at zero (i.e. no location parameter), which prevents the drawback of parameter estimation that defines the support of the variable. Such a condition is necessary to optimally ensure the maximum likelihood estimators (consistency and efficiency; Kendall and Stuart, 1979). This problem will be discussed in more detail in Chapter 4;
- The general form given by $H(.;m,\alpha,\nu)$ with one scale parameter (m) and two shape-parameters (α,ν);
- Flexibility to cover a large range of shapes, thanks to their three parameters;
- Availability of a triplet of sufficient statistics for each of the three distributions of the Halphen family. This warrants an optimal estimation given by the maximum likelihood method.
- In addition, Halphen distributions (HA, HB, HIB) with their limiting forms (Gamma and Inverse Gamma), constitute a complete system.

All these points will be presented in more detail in the following sections. The important advantage of the Halphen distributions is that a sample can be summarized by a three dimensional vector of minimal sufficient statistics. In this aspect the Halphen distributions distinguish themselves from all other three-parameter distributions currently used in hydrology: Log-Pearson type 3, Pearson type 3, three-parameter lognormal, and Generalized Extreme Value. Moreover these last three-parameter distributions, unlike the Halphen system, have a location parameter whose estimation can be troublesome using the maximum likelihood method. The system of maximum likelihood equations for each of the three Halphen distributions are highly nonlinear and must be solved numerically. The parameter estimation methods are presented in Chapter 4.

The Halphen type A (HA) distribution, based on the Bessel function, was first introduced to generalize the Harmonic distribution (Morlat, 1956). The Halphen type B (HB) and Inverse B (HIB), based on the exponential factorial function, were added to form a complete system (similar to the Pearson system, i.e. each sample corresponds to a unique distribution). Morlat (1956) suggested displaying the three Halphen distributions in a diagram (δ_1, δ_2) with $\delta_1 = \ln(A/G)$ and $\delta_2 = \ln(G/H)$ shown as coordinates in Figure 2.1, where A, G, and H are arithmetic, geometric, and harmonic mean, respectively. This diagram is similar to the representation of the Pearson family in the (β_1, β_2) diagram (Bobée and Ashkar, 1991, page 165; Bobée et al. 1993). Where β_1 and β_2 are related to skewness and kurtosis, respectively. By considering moment ratios, the scale parameter m is eliminated, and therefore the value of (δ_1, δ_2) determines the type of Halphen distribution. Figure 2.1 shows the location of Gamma and Inverse Gamma distributions as limiting cases for the different types of Halphen distributions (Bobée et al., 1993). The diagram (δ_1, δ_2) is presented here for illustration purposes and its use will be discussed in more detail in Section 3.5. Indeed, for inference purposes, the sample will be presented by empirical moment ratio and will correspond to the unique distribution of the Halphen system.

Dvorak et al. (1988) showed that the probability density function (pdf) of each of the three Halphen distributions satisfy the following differential equation:

$$\frac{1}{f(x)} \frac{d f(x)}{dx} = \frac{a_0 + a_1 x + a_2 x^2}{x^q} \tag{2.1}$$

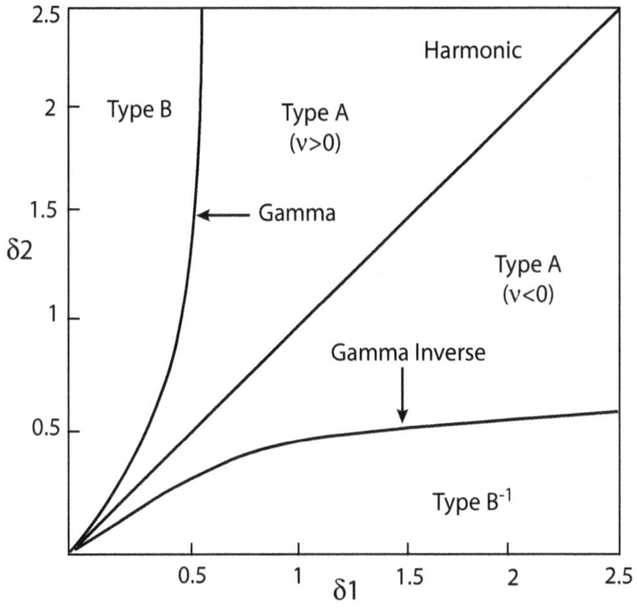

Figure 2.1: (δ_1, δ_2) **-Moment Ratio diagram (From Morlat, 1956).**

Where, the parameters, corresponding to each Halphen distribution, are given in Table 2.1, where m is the scale parameter and α and ν are the shape parameters. Statistical and mathematical properties of the Halphen distributions will be detailed in the subsequent sections.

Table 2.1: Parameters of the differential Eq. (2.1) of each types of the Halphen distributions.

Halphen Distributions	Parameters of the Eq. (2.1)			
	Q	a_0	a_1	a_2
type A	2	αm	$\nu - 1$	$-\alpha / m$
type B	1	$2\nu - 1$	α / m	$-2 / m^2$
type IB	3	$2m^2$	$-\alpha m$	$-(2\nu + 1)$

Figure 2.2 displays different forms of the probability density function of the Halphen distributions. When the mode exists, its explicit expression can be driven from the pdf function and will be given in the following sections.

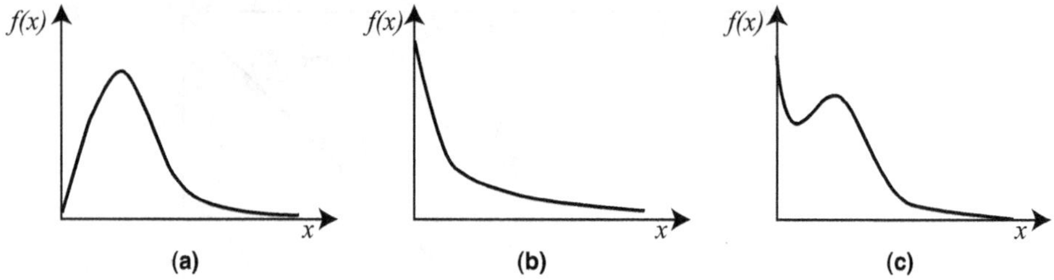

Figure 2.2: Halphen distribution probability distribution forms.

2.2. HALPHEN DISTRIBUTION TYPE A (HA)

2.2.1. Introduction

Halphen (1941) introduced the three parameter type A distribution (HA) which is a generalization of the two parameter Harmonic distribution. The probability density function (pdf) of the HA distribution $HA(x;m,\alpha,v)$ is given by:

$$f_{HA}(x)=\frac{1}{2m^{v}K_{v}(2\alpha)}x^{v-1}\exp\left[-\alpha\left(\frac{x}{m}+\frac{m}{x}\right)\right] \quad , \quad x>0 \tag{2.2}$$

Where $m(>0)$ is the scale parameter, $\alpha(>0)$ and $v\in\mathbb{R}$ are both shape parameters.

$K_v(.)$ is the modified Bessel function of second kind (Watson, 1966) and is given by the following integral:

$$K_v(2\alpha)=\frac{1}{2}\int_0^\infty t^{v-1}\exp\left[-\alpha\left(t+t^{-1}\right)\right]dt= \frac{1}{2m^{v}}\int_0^\infty x^{v-1}\exp\left[-\alpha\left(\frac{x}{m}+\frac{m}{x}\right)\right]dx \tag{2.3}$$

More detail on the modified Bessel function of second kind are given in Appendix A.

The HA distribution has three parameters and offers more flexibility to fit a larger set of data than the two-parameters Harmonic distribution, which is a special case of the HA. So, when the shape parameter v is equal to zero, Eq. (2.2) will correspond to the pdf of the Harmonic distribution $H(x;m,\alpha)$:

$$f_H(x)=\frac{1}{2xK_0(2\alpha)}\exp\left[-\alpha\left(\frac{x}{m}+\frac{m}{x}\right)\right] \quad , \quad x>0 \tag{2.4}$$

Figure (2.3) presents the probability distribution function (fdp) of the HA distribution with four combinations of the shape parameters α and v; the scale parameter m is the same for all the considered cases.

The HA pdf $f_{HA}(x)$ is unimodal with positive skewness and its tail decreases exponentially (Figure 2.2 (a)). More detail on the asymptotic behavior of the HA distribution are given in Chapter 3. The mode is explicitly given by the expression:

$$x_M = m\left[\frac{v-1}{2\alpha} + \sqrt{\left(\frac{v-1}{2\alpha}\right)^2 + 1}\right] \tag{2.5}$$

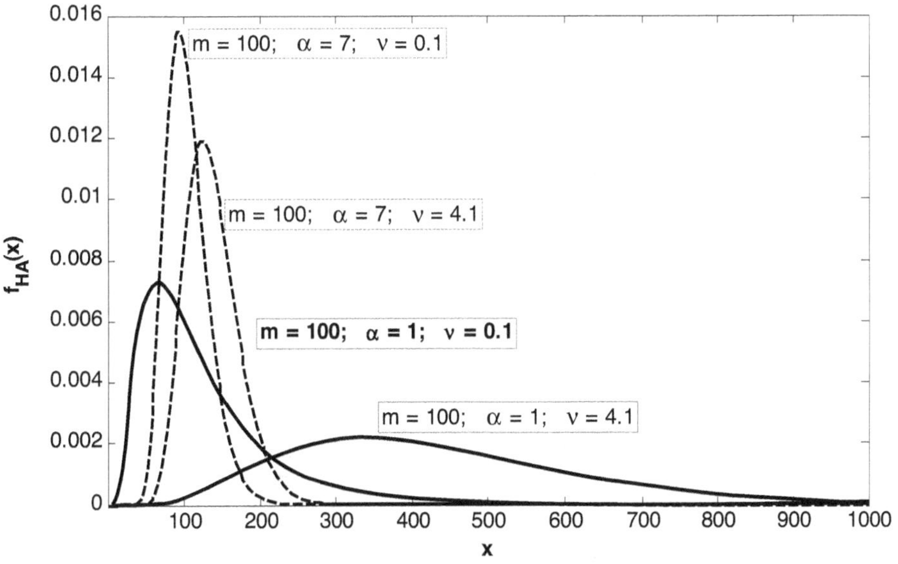

Figure 2.3: Probability density function of the Halphen type A distribution for different sets of parameters.

2.2.2. Moments and Moment Ratios

The non-central moments $\mu'^{(r)}_{HA}$ of the Halphen type A distribution $HA(x;m,\alpha,v)$ can be obtained from the characteristic function $\phi(t)$ (Rice, 2007) given by:

$$\phi_{HA}(t) = \int_{-\infty}^{\infty} e^{itx} dF_{HA}(x) = \int_{0}^{\infty} e^{itx} f_{HA}\left(x\right) dx \tag{2.6}$$

where F_{HA} is the cumulative probability distribution (cdf) corresponding to the pdf f_{HA} given by Eq. (2.2). Indeed, the characteristic function can be represented using the exponential function's power series and then $\mu'^{(r)}_{HA}$, the non-central moment of order r, corresponds to the coefficient of $(it)^r/r!$ in the following development:

$$\phi_{HA}(t)=\sum_{n=0}^{\infty}\frac{(it)^r}{r!}\mu'^{(r)}_{HA} \tag{2.7}$$

Using the expression of the pdf of HA Eq. (2.2), the characteristic function is given by:

$$\phi_{HA}(t)=\frac{1}{2m^v K_v(2\alpha)}\int_0^{\infty} e^{itx}x^{v-1}\exp\left[-\alpha\left(\frac{x}{m}+\frac{m}{x}\right)\right]dx \tag{2.8}$$

By replacing the term e^{itx}, by its Taylor series, the characteristic function is:

$$\phi_A(t)=\frac{1}{2m^v K_v(2\alpha)}\bullet$$

$$\int_0^{\infty}\left[1+\frac{itx}{1!}+\frac{(itx)^2}{2!}+...+\frac{(itx)^r}{r!}+...\right]x^{v-1}\exp\left[-\alpha\left(\frac{x}{m}+\frac{m}{x}\right)\right]dx$$

$$=\frac{1}{2m^v K_v(2\alpha)}\int_0^{\infty}\sum_{r=0}^{\infty}\frac{(it)^r}{r!}x^{v+r-1}\exp\left[-\alpha\left(\frac{x}{m}+\frac{m}{x}\right)\right]dx$$

$$=\frac{1}{2m^v K_v(2\alpha)}\sum_{r=0}^{\infty}\frac{(it)^r}{r!}\int_0^{\infty}x^{v+r-1}\exp\left[-\alpha\left(\frac{x}{m}+\frac{m}{x}\right)\right]dx$$

Using the expression of the function $K_v(2\alpha)$ (Eq. 2.3):

$$\int_0^{\infty}x^{v+r-1}\exp\left[-\alpha\left(\frac{x}{m}+\frac{m}{x}\right)\right]dx=2m^{v+r}K_{v+r}(2\alpha)$$

then,

$$\phi_{HA}(t)=\sum_{r=0}^{\infty}\frac{(it)^r}{n!}\frac{m^r K_{v+r}(2\alpha)}{K_v(2\alpha)} \tag{2.9}$$

We conclude that, for HA distributed random variable X, the rth non-central moment, $\mu'^{(r)}_{HA} = E[X^r]$, is given by the expression:

$$\mu'^{(r)}_{HA} = \frac{m^r K_{v+r}(2\alpha)}{K_v(2\alpha)} \tag{2.10}$$

Non-central moments of the HA distribution exist for all the value of α (> 0), r, and $v \in \mathbb{R}$, and are related to the modified Bessel function of the second kind $K_v(2\alpha)$.

The mean of a random variable X that follows the HA distribution can be deduced from Eq. (2.10) for $r = 1$:

$$\mu'^{(1)}_{HA} = E\left[X\right] = m\frac{K_{v+1}(2\alpha)}{K_v(2\alpha)} = A \tag{2.11}$$

The moments of order –1 and quasi-zero ($\underline{0}$), which correspond to the harmonic and the geometric mean, respectively, are measures of location and are very important to estimate the parameters of the HA distribution. Their expressions are:

$$\mu'^{(-1)}_{HA} = E\left[\frac{1}{X}\right] = \frac{1}{m}\frac{K_{v-1}(2\alpha)}{K_v(2\alpha)} = H^{-1} \tag{2.12}$$

$$\mu'^{(\underline{0})}_{HA} = E\left[\ln X\right] = \ln m + \frac{\partial K_v(2\alpha)/\partial v}{K_v(2\alpha)} = \ln\left(G\right) \tag{2.13}$$

The moment of order –1 can be derived directly from the Eq. (2.10) and its inverse corresponds to the harmonic mean. Detail corresponding to the determination of the moment quasi-zero are given in (Bobée and Ashkar 1991, page 102), and it corresponds to the logarithm of the geometric mean.

Central moments $\mu^{(r)}_{HA}$ can be deduced from Eq. (2.10) through the expression (Kendall and Stuart, 1979):

$$\mu'^{(r)}_{HA} = \sum_{j=0}^{r}\begin{pmatrix} j \\ r \end{pmatrix}\mu'^{(r-j)}_{HA}\left(-\mu'^{(1)}_{HA}\right)^j \tag{2.14}$$

The first central moments are deduced from this expression using Eq. (2.11):

$$\mu'^{(2)}_{HA} = Var\left[X\right] = \frac{m^2}{K_v^2}\left(K_v K_{v+2} - K_{v+1}^2\right) \tag{2.15a}$$

$$\mu'^{(3)}_{HA} = \frac{m^3}{K_v^3}\left(K_{v+3}K_v^2 - 3K_{v+2}K_{v+1}K_v + 2K_{v+1}^3\right) \tag{2.15b}$$

$$\mu'^{(4)}_{HA} = \frac{m^4}{K_v^4}\left(K_v^3 K_{v+4} - 4K_v^2 K_{v+1}K_{v+3} + 6K_v K_{v+1}^2 K_{v+2} - 3K_{v+1}^4\right) \tag{2.15c}$$

For sake of simplicity K_v denotes the Bessel function $K_v(2\alpha)$ (Eq. (2.3) and Appendix A).

The moment ratios, which correspond to the coefficients of variation (C_v), skewness (C_s), and kurtosis (C_k), can be deduced from the last expressions. These coefficients are usually considered to characterize the distributions through diagrams.

The expressions of these coefficients for the HA distributions are:

$$C_V\left(HA\right) = \frac{\sqrt{\mu^{(2)}_{HA}}}{\mu'^{(1)}_{HA}} = \frac{\sqrt{K_v K_{v+2} - K_{v+1}^2}}{K_{v+1}} \tag{2.16a}$$

$$C_s\left(HA\right) = \frac{\mu^{(3)}_{HA}}{\left[\mu^{(2)}_{HA}\right]^{3/2}} = \frac{K_v^2 K_{v+3} - 3K_v K_{v+1} K_{v+2} + 2K_{v+1}^3}{\left(K_v K_{v+2} - K_{v+1}^2\right)^{3/2}} \tag{2.16b}$$

$$C_k\left(HA\right) = \frac{\mu^{(4)}_{HA}}{\left[\mu^{(2)}_{HA}\right]^2} = \frac{K_v^3 K_{v+4} - 4K_v^2 K_{v+1} K_{v+3} + 6K_v K_{v+1}^2 K_{v+2} - 3K_{v+1}^4}{\left(K_v K_{v+2} - K_{v+1}^2\right)^2} \tag{2.16c}$$

In order to simplify, the last expressions, the following notations derived from (Eq. (2.10)) are considered to be:

$$R_A\left(\alpha, v\right) = \frac{K_{v+1}\left(2\alpha\right)}{K_v\left(2\alpha\right)} = \frac{\mu'^{(1)}_{HA}}{m} \tag{2.17}$$

and

$$D_A(\alpha,v)=\frac{K_{v+1}(2\alpha)K_{v-1}(2\alpha)}{K_v^2(2\alpha)}=R_A(\alpha,v)R_A^{-1}(\alpha,v-1) \qquad (2.18)$$

These functions will be used hereafter in Chapter 4 for inference purposes and to present the estimations methods for the Halphen distribution parameters.

The dispersion function, $D_A(\alpha,v)$, is of particular interest for the maximum likelihood estimators of the parameters of the HA distribution. It is worthwhile to note that this function is the product of the moments of order 1 and −1, (Eqs. (2.11) and (2.12)), i.e. the ratio of the arithmetic and the harmonic means.

Some important moments can be re-written as functions of $R_A(\alpha,v)$ and $D_A(\alpha,v)$. Indeed, the arithmetic mean, the harmonic mean, the variance, and the coefficient of variation of the HA distribution are given by:

$$E[X] = mR_A(\alpha,v) \qquad (2.19)$$

$$E[1/X]=\frac{1}{m}R_A^{-1}(\alpha,v-1) \qquad (2.20)$$

$$Var[X]=m^2R_A^2(\alpha,v)[D_A(\alpha,v+1)-1] \qquad (2.21)$$

$$C_V=\sqrt{D_A(\alpha,v+1)-1} \qquad (2.22)$$

Figure 2.4 illustrates the relationship between the coefficient of variation C_V and the coefficient of skewness C_s for the HA distribution with different sets of the shape parameters α (dotted lines) and v (continuous lines). This region is delimited by the curves corresponding to the Gamma and Inverse Gamma distributions (cf. Section 2.2.4).

The Figure 2.4 illustrates the relationship between the coefficient of variation C_V and skewness C_s for the HA distribution with different set of the shape parameters α (continuous lines) and v (dotted lines).

More generally, the moment ratio diagrams serve four purposes: (1) they quantify the proximity between various univariate distributions

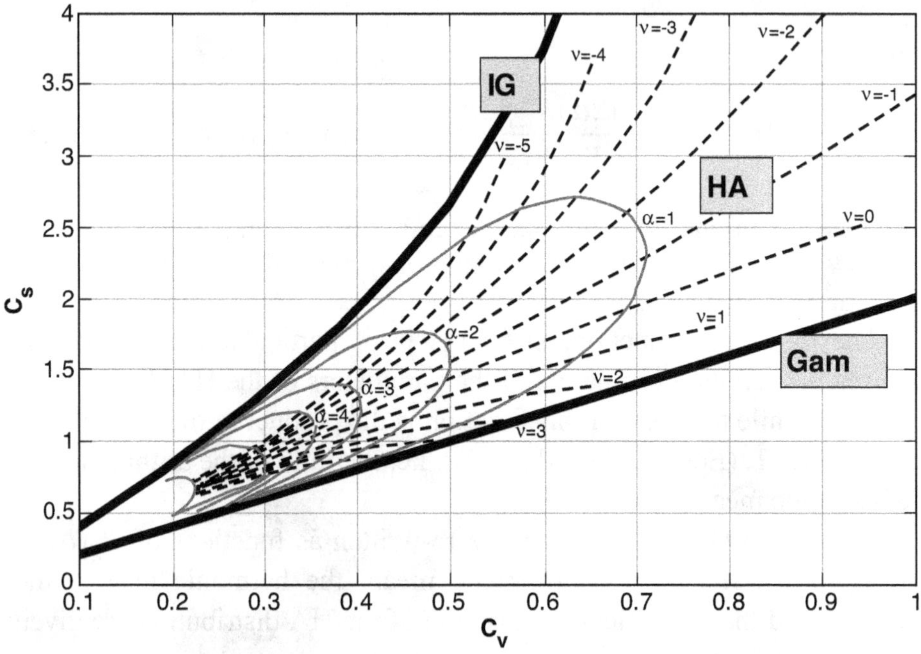

Figure 2.4: HA distribution: Moment ratio diagram (C_v,C_s) for different values of α and ν.

based on their second, third, and fourth moments; (2) they illustrate the versatility of a particular distribution based on the range of values that the various moments can assume; (3) they can be used to create a short list of potential probability models based on a data set for choosing a distribution that models given dataset; and (4) they clarify the limiting relationships between various well-known distribution families. Note that the variance of the empirical estimators of these moments increase with their order.

2.2.3. Sufficient Statistics

An important statistical result is available for the class of the distribution family with three parameters $\theta = (\theta_1, \theta_2, \theta_3)$. The canonical form of the Exponential family has the following density function form:

$$f(x) = \exp\left\{ \sum_{i=1}^{3} c_i(\theta_1, \theta_2, \theta_3) T_i(x) + d(\theta_1, \theta_2, \theta_3) + S(x) \right\} \quad (2.23)$$

The factorization theorem shows that $T(x) = (T_1(x), T_2(x), T_3(x))$ are sufficient statistics for the density function of X. A statistic is *sufficient* if it is just as informative as the full data. The concept was introduced by Fisher (1922) and refined by Neyman (1935). Sufficiency means that the statistic contains just as much information about (some) parameter of the model as the full data. The Halphen type A distribution belongs to the exponential family of continuous probability density functions (pdf), and has three sufficient statistics. Indeed, the HA pdf can be presented as follows:

$$f_{HA}(x) = \frac{1}{2m^v K_v(2\alpha)} x^{v-1} \exp\left[-\alpha\left(\frac{x}{m} + \frac{m}{x}\right)\right]$$
$$= \exp\left\{(v-1)\ln x - \frac{\alpha}{m}x - \alpha m\frac{1}{x} - \ln\left[2m^v K_v\left(2\alpha\right)\right]\right\}$$

The HA pdf has the same form as that of the Eq. (2.23) and the corresponding functions are given by:

- $c_1(m,v,\alpha) = (v-1)$
- $c_2(m,v,\alpha) = \alpha / m$
- $c_3(m,v,\alpha) = \alpha m$
- $d\left(m,v,\alpha\right) = -\ln\left[2m^v K_v\left(2\alpha\right)\right]$
- $T_1\left(x\right) = \ln\left(x\right)$
- $T_2\left(x\right) = x$
- $T_3\left(x\right) = 1/x$
- $S\left(x\right) = 0$

Thus, due to the factorization theorem (Bickel and Doksum, 1977), the statistics $(\ln X, X, 1/X)$ are jointly sufficient for the Halphen type A distribution. Furthermore, since a bijective transformation of a sufficient statistic remains sufficient, we deduce that for a given sample of n independent and identically distributed variables $X_1, X_2, \ldots, X_n$ following HA distribution, the sufficient statistics are given by:

$$T_1\left(\underline{X}\right) = \frac{1}{n}\sum_{i=1}^{n} \ln X_i = \ln G \tag{2.24a}$$

$$T_2\left(\underline{X}\right) = \frac{1}{n}\sum_{i=1}^{n} X_i = A \tag{2.24b}$$

$$T_3\left(\underline{X}\right) = \frac{1}{n}\sum_{i=1}^{n} \frac{1}{X_i} = H^{-1} \tag{2.24c}$$

We conclude that the arithmetic (A), the Harmonic (H), and the Geometric (G) means are sufficient statistics for the HA distribution and thus, are efficient (minimum variance) to estimate the HA parameters.

When E. Halphen (1941) proposed the HA distribution, the objective was to develop a new distribution with lighter tail than the Lognormal distribution (Section 3.3) and which has a sufficient statistics to ensure the existence of efficient estimators of the parameters (see Chapter 4).

2.2.4. Related Distributions

The Gamma (Gam) and Inverse Gamma (IG) distributions can be derived as limiting cases of the HA. Here and in the following parameterization for the Gamma distribution ($Gam(\alpha,\lambda)$) is given by (Bobée and Ashkar 1991):

$$f_{Gam}(x)=\frac{\alpha^{\lambda}}{\Gamma(\lambda)}x^{\lambda-1}\exp[-\alpha x] \quad , \quad x>0 \tag{2.25}$$

Where $\alpha>0$ is the scale parameter and $\lambda>0$ the shape parameter and $\Gamma(\lambda)$ is the Gamma function (Bobée and Ashkar, 1991, Apendix A).

The pdf of the Inverse Gamma, $IG(\alpha,\lambda)$, is defined by

$$f_{IG}(x)=\frac{\alpha^{\lambda}}{\Gamma(\lambda)}\left(\frac{1}{x}\right)^{\lambda+1}\exp\left[-\frac{\alpha}{x}\right] \quad , \quad x>0 \tag{2.26}$$

With the above parameterization, if X has a $Gam(\alpha,\lambda)$ distribution then $Y=1/X$ has an $IG(\alpha,\lambda)$ distribution.

The Generalized Inverse Gaussian distribution (GIG) (Jørgensen 1982, Embrechts 1983, and Seshadri, 1993) corresponds, with a different parameterization, to the HA distribution. Indeed, the pdf of the GIG distribution is given by:

$$f_{GIG}(z)=\frac{\left(\frac{\psi}{\chi}\right)^{v/2}}{2K_{v}\left(\sqrt{\psi\chi}\right)}z^{v-1}\exp\left\{-\frac{1}{2}\left(\frac{\chi}{z}+\psi z\right)\right\} \tag{2.27}$$

If we pose $\alpha=1/2\sqrt{\chi\psi}$ and $m=\sqrt{\chi/\psi}$, Eq. (2.27) corresponds to the pdf of the HA distribution (Eq. (2.2)).

As mentioned in Section 2.2, the Gamma (Gam) and Inverse Gamma (IG) distributions constitute limiting cases of the Halphen distributions. Their pdf are given by Eqs. (2.25) and (2.26), respectively. In fact, these two distributions are the only ones from the Pearson system that satisfy the conditions formulated by Halphen to create a new family of distributions (absence of the location parameter belonging to the Exponential family of distributions). Thus the Gamma and Inverse Gamma distributions have sufficient statistics (Chapter 4).

The next proposition indicates the convergence of the HA distribution to a Gamma (Gam) and Inverse Gamma (IG) distributions.

Proposition 2.1. Let X be a random variable with $X \sim HA(m,\alpha,v)$.

- Suppose that the parameters α and m converge simultaneously to zero with $\alpha / m \rightarrow \omega$ and $\alpha / m \rightarrow \omega$ finite positive numbers. Then, if $v > 0$, $X \sim Gam(\omega = \alpha/m, v)$;

- If α converge to zero and m to infinity such that $\alpha / m \rightarrow 0$ and $\alpha m \rightarrow \theta$ (finite positive number. Then, if $v < 0$, $X \sim IG(\theta = \alpha m, -v)$.

The demonstration of this proposition is available in Perreault et al. (1999a). Note that Sichel distribution corresponds to the discrete version of the HA distribution (Sichel, 1971).

2.3. HALPHEN DISTRIBUTION TYPE B (HB)

2.3.1. Introduction

Following an intensive empirical study of the adequacy of the Halphen type A distribution to many sets of observations, Halphen found it necessary to introduce a new distribution, which has a different asymptotic behavior near zero. This distribution is the Halphen type B (HB).

The probability distribution function of the Halphen type B distribution $HB(y;m,\alpha,v)$ is given:

$$f_{HB}(y) = \frac{2}{m^{2v} ef_v(\alpha)} y^{2v-1} \exp\left[-\left(\frac{y}{m}\right)^2 + \alpha\left(\frac{y}{m}\right)\right] \quad , \quad y > 0 \qquad (2.28)$$

Where m (> 0) is a scale parameter, $\alpha \in \mathbb{R}$ and (v >0) are both shape parameters.

Where for v >0 :

$$ef_v(\alpha) = 2\int_0^{\infty} y^{2v-1} e^{\left(-y^2+\alpha y\right)} dy \tag{2.29}$$

The function $ef_v(\alpha)$ is exponential factorial function (Appendix B) defined by (Halphen 1955) and is related to the Hermite polynomial and the confluent hypergeometric function (Abramovitz and Stegun, 1972). These relationships and other properties of the exponential factorial function are given in Appendix B.

The pdf of the HB distribution has various forms according to the values of the parameters. It can be without mode, have a unique mode or have one mode and an anti-mode (Figure 2.2; Perreault et al. 1999a). These three cases occur in the following conditions.

1. The pdf of the HB distribution admits no mode (Figure 2.2 (b)) if one of the following conditions is verified: $\alpha < 0$ and $v < 0.5$ or $(-\alpha/4)^2 \leq 0.5 - v$;

2. It has one mode (Figure 2.2 (a)) if $\alpha \neq 0$ and $v \geq 0.5$,

3. It possesses one mode and an anti-mode (Figure 2.2 (c)) if $\alpha \leq 0$, $v < 0.5$, and $(-\alpha/4)^2 > 0.5 - v$.

Note that the HB distribution allows the Gamma distribution as limiting case (Section 2.3.4 and Figure 2.1).

2.3.2. Moments and Moment Ratios

The non-central moments $\mu_{HB}'^{(r)}$ of the Halphen type B distribution can be obtained from the characteristic function given by the following expression:

$$\phi_{HB}(t) = \frac{2}{m^{2v} ef_v(\alpha)} \int_0^{\infty} e^{ity} y^{2v-1} \exp\left[-\left(\frac{y}{m}\right)^2 + \alpha\left(\frac{y}{m}\right)\right] dy$$

By replacing the term e^{itx}, by its Taylor series, the characteristic function is:

$$\phi_{HB}(t) = \frac{2}{m^{2v} ef_v(\alpha)} \int_0^\infty \left[\begin{array}{c} 1 + \frac{ity}{1!} + \frac{(ity)^2}{2!} + \ldots \\ + \frac{(ity)^r}{r!} + \ldots \end{array} \right] y^{2v-1} \exp \left[\begin{array}{c} -\left(\frac{y}{m}\right)^2 + \\ \alpha \left(\frac{y}{m}\right) \end{array} \right] dy$$

$$= \frac{2}{m^{2v} ef_v(\alpha)} \int_0^\infty \sum_{r=0}^\infty \frac{(it)^r}{r!} y^{2v+r-1} \exp \left[-\left(\frac{y}{m}\right)^2 + \alpha \left(\frac{y}{m}\right) \right] dx$$

$$= \frac{2}{m^{2v} ef_v(\alpha)} \sum_{r=0}^\infty \frac{(it)^r}{r!} \int_0^\infty y^{2(v+r/2)-1} \exp \left[-\left(\frac{y}{m}\right)^2 + \alpha \left(\frac{y}{m}\right) \right] dy$$

Or by definition the integral in the last equation is proportional to the exponential factorial function

$$ef_{v+r/2}(\alpha) = \int_0^\infty y^{2(v+r/2)-1} \exp \left[-\left(\frac{y}{m}\right)^2 + \alpha \left(\frac{y}{m}\right) \right] dy .$$

Then the characteristic function of the $HB(y;m,\alpha,v)$ distribution is given by (cf. Eq. (2.7)):

$$\phi_{HB}(t) = \sum_{r=0}^\infty \frac{(it)^r}{r!} \frac{m^r ef_{v+r/2}(\alpha)}{ef_v(\alpha)} \tag{2.30}$$

We conclude that for HB distributed random variable *X*, the non-central moment of order *r* of *X*, $\mu_{HB}^{\prime(r)} = E[Y^r]$, are given by the expression:

$$\mu_{HB}^{\prime(r)} = E\left[Y^r\right] = \frac{m^r ef_{v+r/2}(\alpha)}{ef_v(\alpha)} \tag{2.31}$$

Non-central moments of the HB distribution exist for all the value of $\alpha > 0$, only when $v + r/2 > 0 \Rightarrow r > -2v$, since the index of the exponential factorial function should be positive. Then the expression of the first moment (the arithmetic mean) of a random variable *X* that is HB distributed is:

$$\mu_{HB}^{\prime(1)} = E\left[Y\right] = m \frac{ef_{v+1/2}(\alpha)}{ef_v(\alpha)} \tag{2.32}$$

As it will be shown in Section 2.3.3, the non-central moments of order quasi-zero (geometric mean) and order 2, are very important to estimate the parameters of the HB distribution. Their expressions are given by Eqs. (2.33) and (2.34), respectively:

$$\mu_{HB}^{\prime(0)} = E\left[\ln Y\right] = \ln m + \frac{\partial\, ef_{v}(\alpha)/\partial v}{2ef_{v}(\alpha)} \tag{2.33}$$

$$\mu_{HB}^{\prime(2)} = E\left[Y^{2}\right] = \frac{m^{2} ef_{v+1}(\alpha)}{ef_{v}(\alpha)} \tag{2.34}$$

The moment of order 2 can be derived directly from Eq. (2.31). Detail corresponding to the determination of the moment quasi-zero are given in (Bobée and Ashkar, 1991; page 102), and it corresponds to the logarithm of the geometric mean.

In a similar way as for the HA distribution, the central moments $\mu_{HB}^{(r)}$ can be deduced from Eq. (2.14) and Eq. (2.31). The first central moments of the HB distributed random variable are then given by:

$$\mu_{HB}^{(2)} = Var\left[Y\right] = \frac{m^{2}}{ef_{v}^{2}}\left(ef_{v} ef_{v+1} - ef_{v+1/2}^{2}\right) \tag{2.35a}$$

$$\mu_{HB}^{(3)} = \frac{m^{3}}{ef_{v}^{3}}\left(ef_{v+3/2} ef_{v}^{2} - 3ef_{v+1/2} ef_{v+1} ef_{v} + 2ef_{v+1/2}^{3}\right) \tag{2.35b}$$

$$\mu_{HB}^{(4)} = \frac{m^{4}}{ef_{v}^{4}}\begin{pmatrix} ef_{v}^{3} ef_{v+2} - 4ef_{v}^{2} ef_{v+1/2} ef_{v+3/2} + \\ 6ef_{v} ef_{v+1/2}^{2} ef_{v+1} - 3ef_{v+1/2}^{4} \end{pmatrix} \tag{2.35c}$$

For sake of simplicity ef_{v} denote the exponential factorial function $ef_{v}(\alpha)$ (Eq. (2.29)).

The standardized moments which correspond to thecoefficients of variation $\left(C_{V}\right)$, skewness $\left(C_{s}\right)$, and kurtosis $\left(C_{k}\right)$, can be deduced from the last expressions. The expressions of these coefficients for the HB distributions are:

$$C_{V}\left(HB\right) = \frac{\mu_{2}^{1/2}}{\mu_{1}^{\prime}} = \frac{\sqrt{ef_{v} ef_{v+1} - ef_{v+1/2}^{2}}}{ef_{v+1/2}} \tag{2.36a}$$

$$C_s\left(HB\right)=\frac{\mu_3}{\mu_2^{3/2}}=\frac{ef_v^2 ef_{v+3/2}-3ef_v ef_{v+1/2} ef_{v+1}+2ef_{v+1/2}^3}{\left(ef_v ef_{\upsilon+1}-ef_{v+1/2}^2\right)^{3/2}} \tag{2.36b}$$

$$C_k\left(HB\right)=\frac{\mu_4}{\mu_2^2}$$
$$=\frac{ef_v^3 ef_{v+2}-4ef_v^2 ef_{v+1/2} ef_{v+3/2}+6ef_v ef_{v+1/2}^2 ef_{v+1}-3ef_{v+1/2}^4}{\left(ef_v ef_{\upsilon+1}-ef_{v+1/2}^2\right)^2} \tag{2.36c}$$

Note that the form of the HB distribution moments (Eq. (2.31)) is similar to those of the HA (Eq. (2.10)). In order to simplify the expressions, the following notations will be considered for the HB distribution:

$$R_B\left(\alpha,v\right)=\frac{ef_{v+1/2}(\alpha)}{ef_v(\alpha)} \tag{2.37}$$

and

$$D_B\left(\alpha,v\right)=\frac{ef_{v+1}(\alpha)ef_v(\alpha)}{\left[ef_{v+1/2}(\alpha)\right]^2}=R_B\left(\alpha,v+1/2\right)R_B^{-1}\left(\alpha,v\right) \tag{2.38}$$

These functions will be considered in Chapter 4, for the inference purposes, to estimate the parameters of the HB distribution. The $D_B(\alpha,v)$ function is related to the ratio of the non-central moment of order 2 and the square of the mean ($D_B\left(\alpha,v\right)=C_V^2+1$). It constitutes a dispersion measure and therefore is called "dispersion function" of the HB distribution. Some important moments can be re-written as functions of the $R_B(\alpha,v)$ and $D_B(\alpha,v)$ functions. Indeed, the non-central moments of order 1, –1, and 2 of the HB distributed random variable X are given by:

$$E\left[Y\right]=mR_B\left(\alpha,v\right) \tag{2.39}$$

$$E\left[1/Y\right]=\frac{1}{m}R_B^{-1}\left(\alpha,v-1/2\right) \tag{2.40}$$

$$E\left[Y^2\right]=m^2R_B^2\left(\alpha,v\right)D_B\left(\alpha,v\right) \tag{2.41}$$

Then the expressions of the variance and the coefficient of variation can be written as

$$Var\left[Y\right] = m^2 R_B^2(\alpha,v)\left[D_B(\alpha,v)-1\right]$$

and

$$C_V = \sqrt{D_B(\alpha,v)-1}\ .$$

The coefficient of variation C_V of the HB distribution depends only on the dispersion function. The $D_B(\alpha,v)$ function is strictly decreasing with α for fixed value of v, and reaches its maximum when $\alpha \to -\infty$, and, in this case, the HB distribution tends to its limiting case Gamma distribution (cf. Section 2.3.4).

The Figure 2.5 illustrates the relationship between the coefficient of variation C_V and skewness C_s for the HB distribution, with a different set of the shape parameters α (continuous lines) and v (dotted lines). This region is delimited by the curve corresponding to the Gamma distribution (cf. Section 2.3.4).

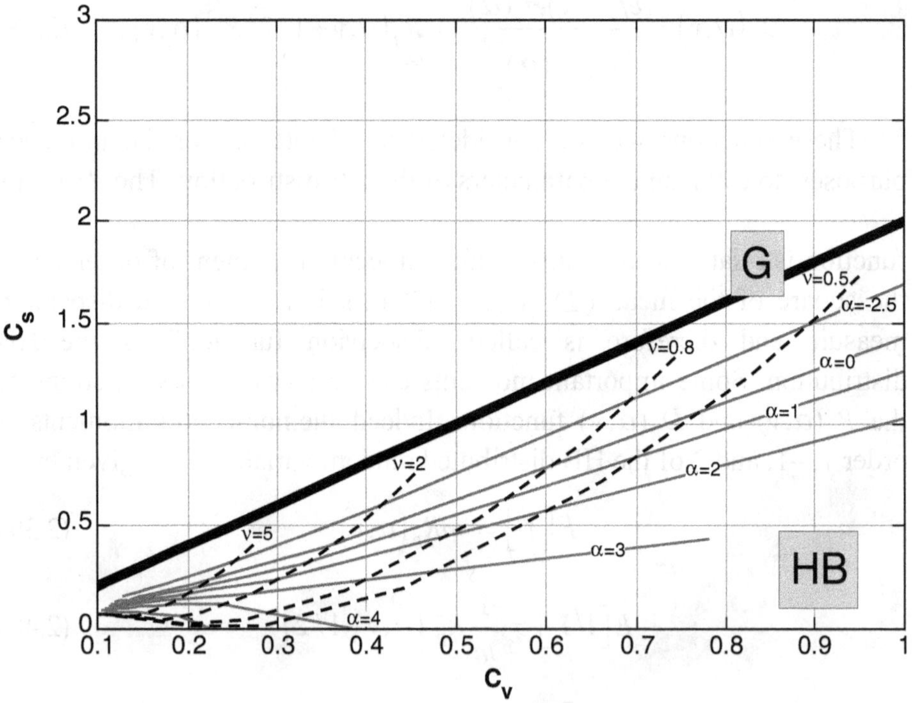

Figure 2.5: HB distribution: The Moment ratio diagram (C_V,C_s) for different values of α and ν.

2.3.3. Sufficient Statistics

Halphen's research was motivated by the development of distributions with sufficient statistics and thus efficient estimators (with minimum variance). The Halphen type B distribution also belongs to the exponential family of continuous probability density functions (Eq. (2.23)), and has three sufficient statistics. So, the pdf of the $HB(y;m,\alpha,v)$ distribution can be presented as follows:

$$f_{HB}(y)=\frac{2}{m^{2v}ef_v(\alpha)}\,y^{2v-1}\exp\left[-\left(\frac{y}{m}\right)^2+\alpha\left(\frac{y}{m}\right)\right]$$

$$=\exp\left\{(2v-1)\ln y-\frac{1}{m^2}y^2+\frac{\alpha}{m}y+\ln\left[2/m^{2v}ef_v(\alpha)\right]\right\}$$

$$=\exp\left\{\sum_{i=1}^{3}c_i(m,v,\alpha)T_i(y)+d(m,v,\alpha)+S(y)\right\}$$

where,

- $c_1(m,v,\alpha)=(2v-1)$
- $c_2(m,v,\alpha)=-1/m^2$
- $c_3(m,v,\alpha)=\alpha/m$
- $d(m,v,\alpha)=\ln\left[2/m^{2v}ef_v(\alpha)\right]$
- $T_1(y)=\ln(y)$
- $T_2(y)=y^2$
- $T_3(y)=y$
- $S(y)=0$

The statistics $(\ln Y, Y, Y^2)$ are sufficient statistics for the Halphen type B (according to the factorization theorem, Bickel and Doksum, 1977). Furthermore, since a bijective transformation of a sufficient statistic remains sufficient, we deduce that for a given sample of n, independent and identically distributed variables $\underline{y}=(y_1, y_2, \ldots, y_n)$. Following HB distribution, the sufficient statistics are:

$$T_1(\underline{y})=\frac{1}{n}\sum_{i=1}^{n}\ln y_i=\ln G \tag{2.42a}$$

$$T_2(\underline{y}) = \frac{1}{n}\sum_{i=1}^{n} y_i^2 = Q \quad (2.42b)$$

$$T_3(\underline{y}) = \frac{1}{n}\sum_{i=1}^{n} y_i = A \quad (2.42c)$$

We conclude that the arithmetic (A), the quadratic (Q), and the Geometric (G) means are sufficient statistics for the HB distribution.

2.3.4. Related Distribution

The next result indicates that the HB distribution has the Gamma distribution as limiting case. Such results are important for parameter estimation and model selection (cf. Chapters 3 and 4).

PROPOSITION 2.2. Let the random variable $X \sim HB(m,\alpha,v)$ and assume that the parameters α and m converge, respectively, to $-\infty$ and $+\infty$, simultaneously, such that $\alpha m \to -\infty$ and $\alpha / m \to -\omega$, where ω is a positive real number. Then, $X \sim Gam(\omega = -\alpha/m, 2v)$.

The demonstration of this proposition is available in Perreault et al. (1999a).

2.4. HALPHEN DISTRIBUTION TYPE INVERSE B (HIB)

2.4.1. Introduction

In this section, we present the third distribution of the Halphen family, Halphen type Inverse B (HIB), sometimes called Halphen type B^{-1}. This distribution has been introduced by Larcher in order to complete the Halphen system (Morlat, 1956).

The probability density function (pdf) of the HIB distribution $HIB(z;m,\alpha,v)$ is given by:

$$f_{HIB}(z) = \frac{2}{m^{-2v} ef_v(\alpha)} z^{-2v-1} \exp\left[-\left(\frac{m}{z}\right)^2 + \alpha\left(\frac{m}{z}\right)\right], \; x > 0 \quad (2.43)$$

where $m(>0)$ is a scale parameter, $\alpha \in \mathbb{R}$ and $(v > 0)$ are both shape parameters and $ef_v(\alpha)$ in (Appendix B), is defined for $v > 0$ and given in Eq. (2.29).

The HIB distribution can be deduced from the HB distribution through an inverse transformation: if Y is HB distributed random variable $[Y \approx F_{HB}(y; m,\alpha,v)]$ then:

$$Z = g(Y) = \frac{1}{Y} \sim HIB\left(z; m^{-1},\alpha,v\right) \tag{2.44}$$

and similarly, if

$$Z \sim HIB\left(z; m',\alpha,v\right)$$

then

$$Y = \frac{1}{Z} \; : \; HB\left(\mathrm{y}; 1/m',\alpha,v\right).$$

Indeed, for any given bijective transformation g, the pdf of the variable Y is related to that of X by the expression: $f_Z(z) = f_Y(g^{-1}(z))J_{g^{-1}}(z)$, where $J_{g^{-1}}(z)$ is the Jacobean of the inverse function g^{-1} evaluated on z. By replacing the expression of the pdf of the HB distribution (Eq. (2.28)), we find:

$$\begin{aligned} f_Z(z) &= f_Z\left(g^{-1}(z)\right)J_{g^{-1}}(z) \\ &= \frac{2}{m^{2v} ef_v(\alpha)}\left(\frac{1}{z}\right)^{2v-1} \exp\left[-\left(\frac{1}{mz}\right)^2 + \alpha\left(\frac{1}{my}\right)\right]\left|-\frac{1}{z^2}\right| \\ &= \frac{1}{2m^{2v} ef_v(2\alpha)} z^{-2v-1} \exp\left[-\left(\frac{m^{-1}}{z}\right)^2 + \alpha\left(\frac{m^{-1}}{z}\right)\right] \quad , \quad z > 0 \end{aligned}$$

The last equation corresponds to the pdf of the HIB distribution Eq. (2.43) with the parameter (m^{-1},α,v).

The HIB distribution has one mode (Figure 2.2a) given by:

$$z_M = m\left[-\frac{\alpha}{4v+2} + \sqrt{\left(\frac{\alpha}{4v+2}\right)^2 + \frac{2}{2v+1}}\right] \tag{2.45}$$

2.4.2. Moments and Moment Ratios

The non-central moments $\mu'^{(r)}_{HIB}$ of the Halphen type Inverse B distribution can not be obtained from the characteristic function as is the case of the HA and HB distribution. Indeed, the characteristic function of the HIB distribution φ_{HIB} is given by

$$\phi_{HIB}(t)=\frac{2}{m^{-2v}ef_v(\alpha)}\int_0^\infty e^{itz}z^{-2v-1}\exp\left[-\left(\frac{m}{z}\right)^2+\alpha\left(\frac{m}{z}\right)\right]dz$$

By replacing the term e^{itz} by its Taylor series, the characteristic function is:

$$\phi_{HIB}(t)=\frac{2}{m^{-2v}ef_v(\alpha)}\bullet$$

$$\int_0^\infty\left[1+\frac{itz}{1!}+\frac{(itz)^2}{2!}+\ldots+\frac{(itz)^r}{r!}+\ldots\right]z^{-2v-1}\exp\left[-\left(\frac{m}{z}\right)^2+\alpha\left(\frac{m}{z}\right)\right]dz$$

$$=\frac{2}{m^{-2v}ef_v(\alpha)}\int_0^\infty\sum_{r=0}^\infty\frac{(it)^r}{r!}z^{-2v+r-1}\exp\left[-\left(\frac{m}{z}\right)^2+\alpha\left(\frac{m}{z}\right)\right]dz$$

$$=\frac{2}{m^{-2v}ef_v(\alpha)}\sum_{r=0}^\infty\frac{(it)^r}{r!}\int_0^\infty z^{-2(v-r/2)-1}\exp\left[-\left(\frac{m}{z}\right)^2+\alpha\left(\frac{m}{z}\right)\right]dz$$

However, this expression exists only when $r<2v$, i.e. when the integral converges. The power series expression of the characteristic function doesn't exist; then we use the integral corresponding definition to find the expression of the non-central moments of order r $\mu'^{(r)}_{HIB}=E\left[Z^r\right]$, of the HIB distributed random variable Z.

$$\mu'^{(r)}_{HIB}=\int_0^\infty z^r f_{HIB}\left(z;m,\alpha,v\right)dz$$

$$=\frac{2}{m^{-2v}ef_v(\alpha)}\int_0^\infty z^{-2(v-r/2)-1}\exp\left[-\left(\frac{m}{z}\right)^2+\alpha\left(\frac{m}{z}\right)\right]dz \qquad (2.46)$$

Given that the integral of the pdf function (Eq. (2.43)) of the HIB distribution is equal to 1, then for $r < 2v$:

$$\int_0^{\infty} z^{-2(v-r/2)-1} \exp\left[-\left(\frac{m}{z}\right)^2 + \alpha\left(\frac{m}{z}\right)\right] dz = \frac{1}{2} m^{-2(v-r/2)} ef_{v-r/2}(\alpha)$$

Thus the non-central moment of the order r of a HIB distributed random variable $[Z \approx F_{IB}(z; m,\alpha,v)]$, is given by:

$$\mu_{HIB}^{\prime(r)} = \frac{m^r ef_{v-r/2}(\alpha)}{ef_v(\alpha)} \tag{2.47}$$

Note: This result can also be deduced for the expression of the non-central moments of the HB distribution (Eq. (2.31)) and the relationship between HIB and HB. Indeed, since

$$Z \approx F_{HIB}(z; m,\alpha,v) \quad \Rightarrow \quad Y = 1/Z \approx F_{HB}(z; m^{-1},\alpha,v),$$

then the moments of these distribution verify:

$$\mu_{HIB}^{\prime(r)}\left(m,\alpha,v\right) = \mu_{HIB}^{\prime(-r)}\left(m^{-1},\alpha,v\right).$$

Using this relationship between the moments of the HIB and HB and Eq. (2.31), the non-central moments of HIB (Eq. (2.47)) can be deduced:

$$\mu_{HIB}^{\prime(r)}\left(m,\alpha,v\right) = \frac{\left(m^{-1}\right)^{-r} ef_{v-r/2}(\alpha)}{ef_v(\alpha)} = \frac{m^r ef_{v-r/2}(\alpha)}{ef_v(\alpha)}.$$

Since the index of the exponential factorial function should be positive, these moments exist only for the orders r that verify the condition $r < [2v]$. From the general expression (Eq. (2.47)), the mean of a HIB distributed random variable $Z \approx F_{HIB}(z; m,\alpha,v)$ is given by:

$$\mu_{HIB}^{\prime(1)} = E\left[Z\right] = m\frac{ef_{v-1/2}(\alpha)}{ef_v(\alpha)} \tag{2.48}$$

The principal first central moments of the HIB distribution can be deduced from the Eq. (2.14) and the Eq. (2.47) and are given by:

$$\mu_{HIB}^{(2)} = Var\left[Z\right] = \frac{m^2}{ef_v^2}\left(ef_{v-1} ef_v - ef_{v-1/2}^2\right) \tag{2.49a}$$

$$\mu_{HIB}^{(3)} = \frac{m^3}{ef_v^3}\left(ef_{v-3/2}ef_v^2 - 3ef_{v-1}ef_v ef_{v-1/2} + 2ef_{v-1/2}^3\right) \tag{2.49b}$$

$$\mu_{HIB}^{(4)} = \frac{m^4}{ef_v^4}\begin{pmatrix} ef_{v-2}ef_v^3 - 4ef_{v-3/2}ef_{v-1/2}ef_v^2 + \\ 6ef_{v-1}ef_{v-1/2}^2 ef_v - 3ef_{v-1/2}^4 \end{pmatrix} \tag{2.49c}$$

For sake of simplicity ef_v corresponds to $ef_v(\alpha)$. The moment ratios, which correspond to the coefficients of variation (C_V), skewness (C_s), and kurtosis (C_k), can be deduced from the last expressions. The expressions of these coefficients for the HIB distributions are:

$$C_V(HIB) = \frac{\sqrt{\mu_{HIB}^{(2)}}}{\mu_{HIB}'^{1}} = \frac{\sqrt{ef_{v-1}ef_v - ef_{v-1/2}^2}}{ef_{v-1/2}} \tag{2.50a}$$

$$\begin{aligned} C_s(HIB) &= \frac{\mu_{HIB}^{(3)}}{\left[\mu_{HIB}^{2}\right]^{3/2}} \\ &= \frac{ef_{v-3/2}ef_v^2 - 3ef_{v-1}ef_{v-1/2}ef_v + 2ef_{v-1/2}^3}{\left(ef_{v-1}ef_v - ef_{v-1/2}^2\right)^{3/2}} \end{aligned} \tag{2.50b}$$

$$\begin{aligned} C_k(HIB) &= \frac{\mu_{HIB}^{(4)}}{\left[\mu_{HIB}^{(2)}\right]^2} \\ &= \frac{ef_{\upsilon-2}ef_\upsilon^3 - 4ef_{\upsilon-3/2}ef_{\upsilon-1/2}ef_\upsilon^2 + 6ef_{\upsilon-1}ef_{\upsilon-1/2}^2 ef_\upsilon - 3ef_{\upsilon-1/2}^4}{\left(ef_{\upsilon-1}ef_\upsilon - ef_{\upsilon-1/2}^2\right)^2} \end{aligned} \tag{2.50c}$$

Given the relationship between HB and HIB distributions, the $D_B(\alpha,v)$ and $R_B(\alpha,v)$ functions Eqs. (2.37) and (2.38), respectively, play an important role in the parameter estimation procedures. Some important moments of the HIB distribution can be re-written as functions of $R_B(\alpha,v)$ and $D_B(\alpha,v)$.

$$E\left[Z\right] = mR_B^{-1}\left(\alpha, v-1/2\right) \tag{2.51}$$

$$E\left[1/Z\right] = \frac{1}{m}R_B\left(\alpha, v\right) \tag{2.52}$$

$$E\left[\frac{1}{Z^2}\right]=\frac{1}{m^2}R_B^2(\alpha,v)D_B(\alpha,v) \tag{2.53}$$

$$Var[Z]=m^2R_B^{-2}(\alpha,v-1/2)\left[D_B(\alpha,v-1)-1\right] \tag{2.54}$$

$$C_V=\sqrt{D_B(\alpha,v-1)-1} \tag{2.55}$$

As shown for the HB distribution, the coefficient of variation C_V of the HIB distribution depends only on the dispersion function. The $D_B(\alpha,v)$ function is strictly decreasing on α for fixed value of v, and reaches its maximum when $\alpha \to -\infty$, and in this case, the HIB distribution tends to its limiting case Inverse Gamma distribution (see Section 2.4.4).

The Figure 2.6 displays the relationship between the coefficient of variation (C_V) and skewness (C_s) for the HIB distribution, with a different set of the shape parameters α (continuous lines) and v (dotted lines). Note that the HIB distribution can cover the part of the diagram with relatively low coefficient of variation and with high skewness. This region is delimited by the curve corresponding to the Inverse Gamma distribution (cf. Section 2.4.4).

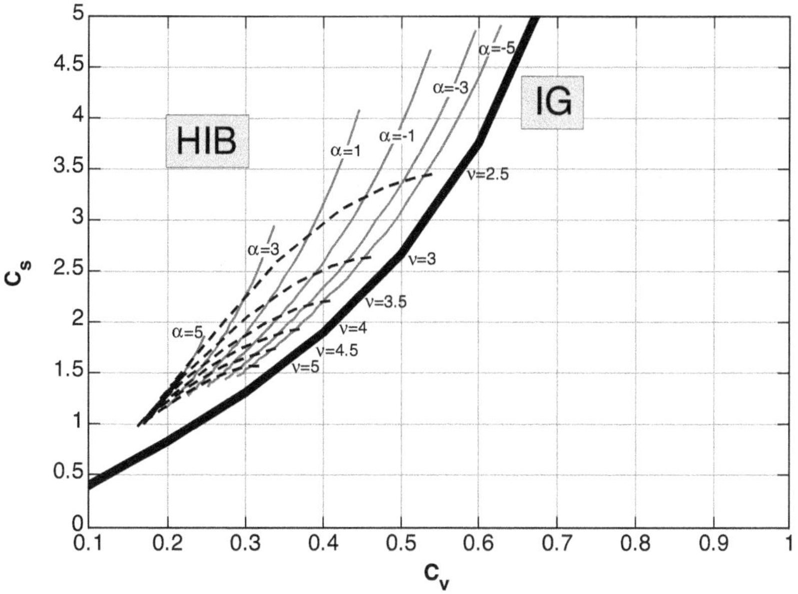

Figure 2.6: HIB distribution: Moment ratio diagram (C_V,C_s) for different values of α and v.

2.4.3. Sufficient Statistics

Halphen's research was motivated by the availability of several probability distribution shapes, especially for the right tail (extreme values); and the sufficiency of the parameter estimators. The Halphen type Inverse B distribution (HIB) belongs also to the exponential family of continuous probability density functions, and has three jointly sufficient statistics. Indeed, the pdf can be presented as in Eq. (2.23):

$$f_{HIB}(z)=\frac{2}{m^{-2v}ef_v(\alpha)}z^{-2v-1}\exp\left[-\left(\frac{m}{z}\right)^2+\alpha\left(\frac{m}{z}\right)\right]$$

$$=\exp\left\{-(2v+1)\ln z-m^2\frac{1}{z^2}+\alpha m\frac{1}{z}+\ln\left[\left(\frac{2}{m^{-2v}}ef_v(\alpha)\right)\right]\right\}$$

$$=\exp\left\{\sum_{i=1}^{3}c_i(m,v,\alpha)T_i(z)+d(m,v,\alpha)+S(z)\right\}$$

where,

- $c_1(m,v,\alpha)=-(2v+1)$
- $c_2(m,v,\alpha)=-m^2$
- $c_3(m,v,\alpha)=\alpha m$
- $d(m,v,\alpha)=\ln\left[2/m^{-2v}ef_v(\alpha)\right]$
- $T_1(z)=\ln(z)$
- $T_2(z)=1/z^2$
- $T_3(z)=1/z$
- $S(z)=0$

The statistics $\ln Z, Z^{-1}, Z^{-2}$ are sufficient statistics for the Halphen type Inverse B (according to the factorization theorem, Bickel and Doksum, 1977). Furthermore, since a bijective function of a sufficient statistic remains sufficient, we deduce that for a given sample of n independent and identically distributed variables $\underline{z}=(z_1,z_2,\ldots,z_n)$ following HIB distribution, the sufficient statistics are:

$$T_1(\underline{z})=\frac{1}{n}\sum_{i=1}^{n}\ln z_i=\ln G \qquad (2.56a)$$

$$T_2(\underline{z})=\frac{1}{n}\sum_{i=1}^{n}\frac{1}{z_i^2}=QI^{-1} \qquad (2.56b)$$

$$T_3(\underline{z})=\frac{1}{n}\sum_{i=1}^{n}\frac{1}{z_i}=H^{-1} \qquad (2.56c)$$

We conclude that the Geometric (G), the Harmonic (H), and the Inverse Quadratic (IQ) means are sufficient statistics for the HIB distribution.

Note: Given the relationship between the HB and HIB distributions (Eq. (2.44)), the last result could be deduced from the sufficient statistics of the HB distribution (The Geometric, the Arithmetic, and Quadratic means).

2.4.4. Related Distribution

As shown for the HA (Figure 2.4) with Gamma and Inverse Gamma distribution also for HB (Figure 2.5) with Gamma distribution, the HIB distribution (Figure 2.6) has the Inverse Gamma distribution as limiting case. Such result can be deduced from the relationship between HB and HIB which corresponds to the same transformation linking Gamma to Inverse Gamma.

Proposition 2.3. Let the random variable $X \sim HIB(m,\alpha,v)$ and assume that the parameters α and m converge, respectively, to $-\infty$ and $+\infty$, simultaneously, such that $\alpha m \to -\infty$ and $\alpha / m \to -\theta$, where ω is a positive real number. Then, $X \sim IG(\theta = -\alpha m, 2v)$.

The demonstration of this proposition is available in Perreault et al. (1999a).

2.5. SUMMARY

In this chapter we presented mathematical and statistical properties of the Halphen distributions. The main properties concern: the absence of location parameter; the existence of sufficient statistics and the availability of a complete system to represent different shapes of the frequency distributions. Table 2.2 summarizes the sufficient statistics of the three distributions and correspond to the moments of order:

- -1 (H^{-1}), $\underline{0}$ (quasi-zero, $\ln G$) and 1 (A) for HA;
- $\underline{0}$ (quasi-zero, $\ln G$), 1 (A) and 2 (Q) for HB;
- -2 (IQ^{-1}), -1 (H^{-1}) and $\underline{0}$ (quasi-zero, , $\ln G$) for HIB.

The availability of such statistics ensures the optimality of the maximum likelihood estimators. Note also that the order of these statistics is lower than the order of the classical moment method.

Table 2.2: Sufficient statistics of the Halphen distributions.

Halphen distributions	Sufficient statistics		
Type A (Eq. (2.24))	H^{-1}	$\ln G$	A
Type B (Eq. (2.42))	$\ln G$	A	Q
Type HIB (Eq. (2.56))	IQ^{-1}	H^{-1}	$\ln G$

The idea of complete system is clearly illustrated through the moment ratio diagrams presented in Figure 2.1 and already discussed in Section 2.1. In addition we may consider the (C_V, C_s) ratio diagram for each Halphen distribution. For HA shows that, this distribution can be considered for moderate skewness and variation (Figure 2.4). The HB distribution allows representing high variations with small skewness (Figure 2.5) and conversely for HIB (i.e. small variation coefficient with high skewness, Figure 2.6).

It is possible to put together the three figures on only one figure to illustrate again the notion of complete system. Figure 2.7 presents the (C_V, C_s) ratio diagram for the Halphen system with their limiting distributions Gamma and Inverse Gamma.

The three regions cover the (C_V, C_s) plan and are limited by the distributions Gamma between HA and HB, and the Inverse Gamma for HA with HIB. This diagram in Figure 2.7 is similar to the delta diagram in Figure 2.1, in the sense that they illustrate how complementary these three distributions are when representing the dataset of independent and identically distributed observations. In addition, the delta diagram (δ_1, δ_2) is based on sufficient statistics and the related estimators have minimum variance.

The following points summarize the principal characteristics of the Halphen system:

- Its flexibility to cover a large range of shapes, thanks to their three parameters (Figures 2.1 and 2.7);
- the availability of a triplet of sufficient statistics for each of the three distributions (Table 2.2); as they belong to the Exponential distribution family (Eq. (2.23)) due to the absence of a location parameter;

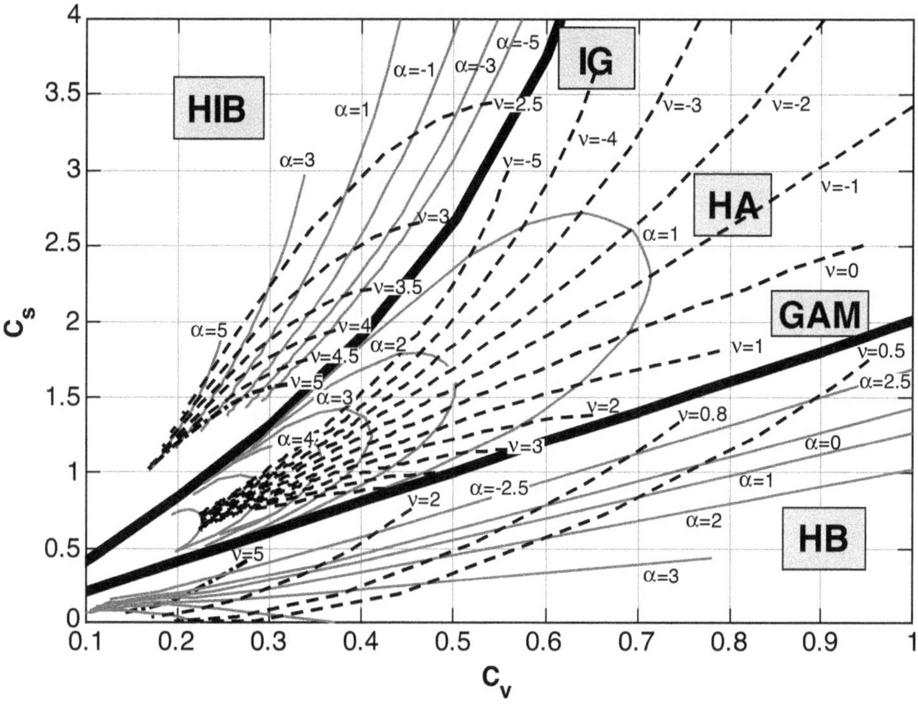

Figure 2.7: Moment ratio diagram (C_V,C_s) for the Halphen system.

- the last properties ensure the optimality of the maximum likelihood estimators (cf. Chapter 4) and,

- the existence of a complete system formed by the three Halphen distributions (HA, HB, HIB) with their limiting forms (Gamma and Inverse Gamma). Indeed, for any sample, the corresponding point (δ_1, δ_2) in Figure 2.1 or Figure 2.7 for the (C_V, C_s), is associated to one and only one member of the Halphen family or their limiting cases.

The mathematical properties, presented in this chapter, will be considered, further, to study the asymptotic behavior of the Halphen distributions. Indeed, to model extremes, the shape of the probability density function is important to select the most adequate distribution. Statistical properties are needed to study the behavior of the parameter estimators and derive the most efficient approaches to perform statistical inference with the Halphen distributions. These main parts, for model comparison based on tail behavior and estimator property assessment, will be presented and discussed in more detail in Chapters 3 and 4, respectively.

Chapter 3
Asymptotic Behavior of the Halphen Distributions

3.1. INTRODUCTION

In the Frequency Analysis (FA) the selection of the most adequate fit is important to obtain robust estimates of extreme quantiles, especially for large return period. This procedure is related to extreme value theory (EVT), which is often derived from asymptotic properties according to the Fisher-Tippet theorem (Fisher and Tippet, 1928). Conventional estimates of flood exceedance quantiles are highly dependent on the form of the underlying flood frequency distribution, especially on the form of the right tail.

In this chapter, the asymptotic behavior of the Halphen distributions is investigated. A comparative study with the Generalized Extreme Value distribution is presented to assess the properties of the large return period quantile estimators.

3.2. CLASSES OF HEAVY TAILED DISTRIBUTIONS AND EXTREME VALUE THEORY

Several definitions can be attributed to heavy tailed distributions, called also fat-tailed, thick-tailed, or long-tailed. A frequently used definition of heavy-tailed distributions is based on the 4th central moment. If X is a random variable and μ_X and σ_X are, respectively, the mean and the standard deviation, then the distribution of X is called heavy-tailed if

$$C_k = E\left[\left(\frac{X-\mu_X}{\sigma_X}\right)^4\right] > 3 \qquad (3.1)$$

$(C_k - 3)$ is known as the excess kurtosis because the 4th central moment (kurtosis) of the Normal distribution is 3. However, in hydrology, the term heavy tailed is used for distributions which have a restriction with respect to moment existence. The difference between the Normal and the heavy tailed distribution was illustrated by Bendjoudi and Hubert (1998) who showed that the event of return period $T = 1000$ estimated by the Normal distribution, corresponds to T=100 years when the HIB distribution (which is heavy tailed) is a suitable fit to the precipitation data series. Figure 3.1 shows that heavy tailed distribution quantiles are larger than those of the Normal distribution (especially for a large non-exceedance probability), even if their first two moments (mean and variance) are similar. This comparison is illustrated using:

- the Halphen type Inverse B distribution (with parameters $\nu = 3$, $\alpha = 3.2$, and $m = 90$;
- the corresponding Exponential Factorial function :

  $ef_{\nu=3}(\alpha = 3.2) = 1695.13,$

  $ef_{\nu=2.5}(\alpha = 3.2) = 687.06,$

  $ef_{\nu=2}(\alpha = 3.2) = 297.92$

  and Eqs. (2.51) and (2.54);
- the Normal distribution with the same mean $E[X] = 110$ and variance $Var[X] = 80$.

Even if these distributions do not have the same parameter space dimension, this comparison illustrates the difference concerning their tail behavior. Indeed, the parameter space dimension affects mainly the variance of the estimated quantiles and not their tail behaviors.

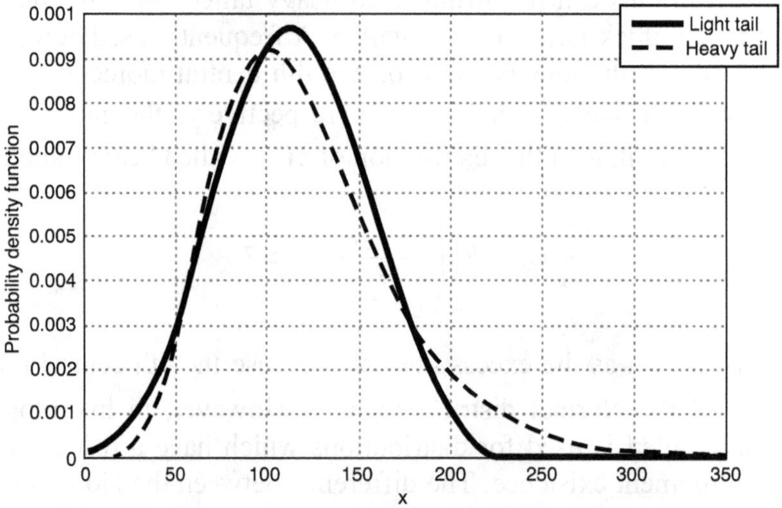

Figure 3.1: Illustration of the difference between a heavy tailed distribution (Halphen type Inverse B) HIB(90;3.2;3) and light tailed (Normal distribution N(110;80)).

Note that the classification based on Eq. (3.1) is vague and cannot be used for distributions with an infinite 4th moment. One tail ranking can be obtained for particular classes of distributions. These classes of distributions are nested ($A \subset B \subset C \subset D \subset E$) and can be presented as in Figure 3.2 (Werner and Upper, 2002, and El Adlouni and Bobée, 2010):

E: distributions with non-existence of exponential moments

D: sub-exponential distributions

C: regularly varying distributions

B: Pareto type tail distributions

A: α-stable (non-normal) distributions

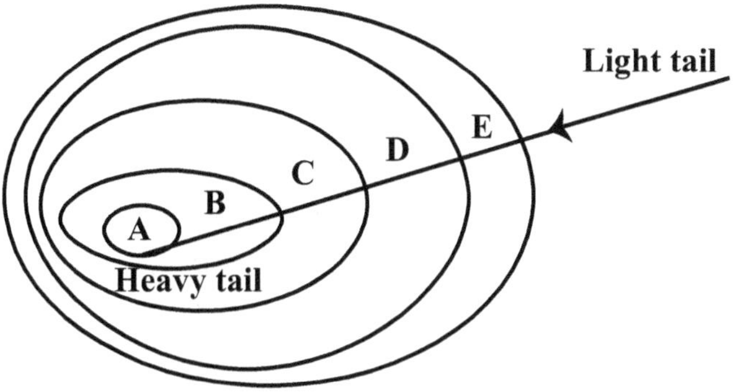

Figure 3.2: Classes of heavy tailed distributions (from Werner and Upper, 2002).

Classes C and B are very important considering their connection to classical extreme value theory (EVT). A main topic of EVT is the modeling of the fluctuation of the maxima dataset. Let $Y_1;Y_2;\ldots;Y_p$ be a sequence of ***independent and identically distributed*** random variables (Berman, 1962), then the sample maximum X is defined as $X = \max(Y_1;Y_2;\ldots;Y_p)$. For example, for the maximum annual flow *X*, *Y* is the daily stream flow and $p = 365$. The Fisher-Tippett theorem (Fisher and Tippett, 1928) states that if the properly normalized sample maximum converges, when p is large, to a non-degenerate distribution, then it belongs to one of the following three distributions (cumulative distribution function, cdf):

Gumbel (EV1): $$\Lambda(x) = \exp\left(-e^{-x}\right) \qquad x \in \mathbb{R} \qquad (3.2)$$

Fréchet (EV2): $$\Phi_\xi(x)=\begin{cases}0 & x\le 0\\ \exp\left(-x^{-\xi}\right) & x>0\end{cases}\qquad \xi>0 \tag{3.3}$$

reverse Weibull (EV3): $$\Psi_\xi(x)=\begin{cases}exp\left(-\left(-x^{-\xi}\right)\right) & x\le 0\\ 0 & x>0\end{cases}\qquad \xi<0 \tag{3.4}$$

In this case, the distribution of the original variable Y, belongs to the Max-Domain of Attraction (MDA) of Λ, Φ_ξ or Ψ_ξ. Von Mises (1954) proposes Generalized Extreme Value (GEV) distribution, which includes the three limit distributions. Another re-parameterization given by Jenkinson (1955), with $\kappa=-\xi$, is generally used in hydrology and is given by:

$$F_{GEV}(x)=\begin{cases}\exp\left[-\left(1-\dfrac{\kappa}{\alpha}(x-u)\right)^{1/\kappa}\right] & \kappa\neq 0\quad \text{(EV2 and EV3)}\\ \exp\left[-\exp\left(-\dfrac{(x-u)}{\alpha}\right)\right] & \kappa=0\quad \text{(EV1)}\end{cases} \tag{3.5}$$

whenever $1-\kappa((x-u)/\alpha)\ge 0$ (to define the support). When the shape parameter κ is positive (respectively negative), the GEV corresponds to the reverse Weibull (EV3) (respectively Fréchet, EV2) distribution. The limiting case, $\kappa=0$, corresponds to the Gumbel (EV1) distribution. A comparison of the EV1 and EV2 distributions, potential candidates for modeling extreme data in hydrology, is given by Bernier (1959) and is based on the asymptotic behavior of classified discharge curves. The simplicity of the GEV distribution function leads to an explicit expression of the cumulative distribution function to compute quantile for a given non-exceedance probability q and is given by:

$$x_T=u+\frac{\alpha}{\kappa}\left[1-\left(\log\left(1-\frac{1}{T}\right)\right)^{\kappa}\right] \tag{3.6}$$

For a given return period $T=1/1-q$ where q is the probability of non-exceedance.

This explicit quantile formula is also useful to generate GEV samples by inversion method based on the generation of uniform distribution. In practice, in order to identify the appropriate form of the GEV distribution (EV1, EV2, or EV3), the sign of the shape parameter should be determined. Therefore, the efficiency of the estimation procedure of the GEV parameters is of a great importance. However, when the shape parameter κ is non-null, which is the case for EV2 and EV3, the support of the random variable depends on the parameters and the pdf of the GEV distribution cannot be presented as a member of the exponential family (Eq. (2.23)). This problem generally leads to estimators with very high variance (Smith, 1985) and will be treated in more detail in Chapter 4. Bernier (1956) presents one of the first studies on the use of EVT in hydrology and shows that the convergence to the limiting distribution can be very slow and depends on the statistical characteristics of the distribution tail. Indeed, when the initial variable Y, is Normal (asymptotic behavior equivalent to $\sqrt{\log T}$), the convergence of the corresponding maxima variable X, to the Gumbel limiting distribution, will be slower when compared to other distributions such as the Pearson type 3 (asymptotic behavior equivalent to $\log T$; cf. Table 3.1; Section 3.3.1).

Based on the previous EVT result, distributions can be classified into three categories. The first one corresponding to EV1 (MDA of Λ) and contains distributions with medium tails. The second one is the class of heavy tailed distributions, whose extremes follow the EV2 distribution (the MDA of Φ_ξ). The third one is the class of short-tailed distributions with finite end-point and contains the distributions of MDA of Ψ_ξ. The class of distributions that belong to the MDA Φ_ξ (Fréchet distribution) is very important when studying heavy tailed behavior. Distributions of this class have a regularly varying tail, i.e. belonging to Class C. Note that the heaviness of the tails depends negatively on the tail index ξ (given in Eq. (3.3)).

Applications of the EVT in hydrology are restricted by the length of the dataset and serial correlation structure. In the case of flood dataset:

- daily flow values Y_i are not statistically independent and identically distributed and;
- the length of the data series ($p = 365$) is not large enough to apply asymptotic results.

Thus rigorously, the use of the GEV distribution has no strong theoretical basis to represent hydrological extreme value datasets. However, it can be considered as a candidate to fit extremes as well any other distributions. This justifies the approach used in FFA based on the fit of several probability distributions and the use of criteria to select the best fit. The classification presented here has the objective to discriminate between distributions with respect to their right tail and will be helpful for best fit selection. This will be discussed thoroughly in Section (3.3).

3.3. ASYMPTOTIC BEHAVIOR AND DECISION SUPPORT SYSTEM

3.3.1. Asymptotic Behavior

This section, we present in more detail the tail behavior of the Halphen distributions. The asymptotic behavior of the Halphen distribution has been studied by Dvorak et al. (1988); Ouarda et al. (1994); Perreault et al. (1999b); El Adlouni et al. (2008); and El Adlouni and Bobée (2010). For a given return period T, that is large enough, the following equivalence holds (Gumbel, 1958, Ouarda et al. 1994):

$$T \underset{x_T \to \infty}{:} \left(\frac{1}{f(x_T)} \right)' = -\frac{f'(x_T)}{f^2(x_T)} \tag{3.7}$$

Indeed, for a large value of T, the probability of exceedance function $\bar{F}(x_T) = 1 - F(x_T) = 1/T = p$ and the pdf $f(x_T)$, both converge to zero.

The Hôpital's rule states that, for large T:

$$\lim_{x_T \to \infty} \frac{f(x_T)}{1 - F(x_T)} \simeq \lim_{x_T \to \infty} \frac{f'(x_T)}{-f(x_T)} \tag{3.8}$$

$$\Rightarrow \qquad T = \frac{1}{1 - F(x_T)} \underset{x_T \to \infty}{\sim} -\frac{f'(x_T)}{f^2(x_T)} = \left(\frac{1}{f(x_T)} \right)' \tag{3.9}$$

In the following section, the functions representing the asymptotic behavior of the Halphen distributions are developed and compared to some distributions usually used in frequency analysis. In order to determine the asymptotic behavior of the Halphen distribution, the equivalence, given by Eq. (3.9), is used. The asymptotic behavior of distributions, commonly used in hydrology, including the Halphen distributions with their limiting cases, is presented in Table 3.1. Four groups are deduced from this classification. The first one (Type I) contains distributions for which the tail is a power function of the return period. The second class (Type II) contains the Lognormal distribution with tail behavior similar to that of the power law. The tail of the distributions in Type III is a power function of the logarithm of the return period and almost all these distributions belong to the Class D of sub-exponential distributions. The last class (Type IV) contains distributions with an upper bounded support.

By combining these two classifications, distributions commonly used in hydrology can be put in order with respect to their tails. Note that almost all these distributions are available in the software HYFRAN-PLUS (CHS, 2002).

Figure 3.3 presents sub-exponential (Class D), regularly varying (Class C) and stable distributions (Class A) ordered from light tailed (from the left) to heavy tailed (to the right) distributions (Figure 3.1). The distributions of each class are in the upper quares and the limiting cases in down squares.

This classification emphasizes the need to develop techniques to discriminate between Class C, of regularly varying distributions, and Class D (Type III), of the other sub-exponential distributions. This is especially true between Class C (Type I with asymptotically power type tail) and the Lognormal distribution. The generation processes of the Lognormal and power-law distributions are often very connected. This phenomenon has been studied in social sciences, biology, and computer sciences. A useful discussion can be found in Reed (2002), Reed and Hughes (2002), and Mitzenmacher (2004). It turns out that many seemingly small effects in the generative process of the Lognormal distribution can lead to a power-law tail instead (Champernowne 1953, Mandelbrot 1997, 2003, and Turcotte 1997). Generally, the Exponential distribution is not considered for extreme frequency analysis, it is considered here as reference behavior.

Table 3.1: Asymptotic behavior classification of the commonly used distributions in hydrology (from Ouarda et al. 1994).

Class	Characteristics	Distribution	Parameters
Type I: $x \approx T^P$			
Class C	$P = 1/\alpha$	Log-Pearson 3 (α,λ,m)	$\alpha > 0,\ \lambda > 0,\ m \in \mathbb{R}$
Class C	$P = 1/\alpha$	Log-Logistic (α,λ)	$\alpha > 0,\ \lambda > 0$
Class C	$P = -1/s\lambda$	Generalized Gamma (s,α,λ)	$s < 0,\ \alpha > 0,\ \lambda > 0$
Class C	$P = 1/\lambda$	Inverse Gamma (α,λ)	$\alpha > 0,\ \lambda > 0$
Class C	$P = -1/k$	Fréchet (α,k,u)	$\alpha > 0,\ k > 0,\ u \in \mathbb{R}$
Class B	$P = -1/k$	Generalized Pareto (α,k)	$\alpha > 0,\ k < 0$
Class C	$P = 1/2\nu$	Halphen type Inverse B (α,ν,m)	$\alpha \in \mathbb{R},\ \nu > 0,\ m > 0$
Type II: $x \approx \exp[(\ln T)^{1/2}$		Lognormal 2 (μ,σ)	$\mu \in \mathbb{R},\ \sigma > 0$
------		Lognormal 3(μ,σ,m)	$\mu \in \mathbb{R},\ \sigma > 0,\ m \in \mathbb{R}$
Type III: $x \approx (\ln T)^P$	$P = 1$	Pearson type 3 (α,λ,m)	$\alpha > 0,\ \lambda > 0,\ m \in \mathbb{R}$
Class D			
Class D		Gamma (α,λ)	$\alpha > 0,\ \lambda > 0$
Class E		Exponential (α,m)	$\alpha > 0,\ m \in \mathbb{R}$
Class D		Halphen type A (α,ν,m)	$\alpha > 0,\ \nu \in \mathbb{R},\ m > 0$
Class D		Log-F(λ,β)	$\lambda > 0,\ \beta > 0$
Class D		Gumbel (α,u)	$\alpha > 0,\ u \in \mathbb{R}$
Class D	$P = 1/2$	Halphen type B(α,ν,m)	$\alpha \in \mathbb{R},\ \nu > 0,\ m > 0$
Class D	$P = 1/s$	Generalized Gamma (s,α,λ)	$s > 0,\ \alpha > 0,\ \lambda > 0$
Class D	$P = 1/c$	Reverse Weibull (α,c)	$\alpha > 0,\ c > 0$
Type IV: $x \le P$	$P = m$	Pearson 3(α,λ,m)	$\alpha < 0,\ \lambda > 0,\ m \in \mathbb{R}$
(Upper bounded support)	$P = 0$	Gamma (α,λ)	$\alpha < 0,\ \lambda > 0$
	$P = \exp(m/\ln_a(e))$	Log-Pearson 3(α,λ,m)	$\alpha < 0,\ \lambda > 0,\ m \in \mathbb{R}$
	$P = 1$	Log-Logistic (α,λ)	$\alpha < 0,\ \lambda > 0$
	$P = u+\alpha/k$	Weibull 3 (α,k,u)	$\alpha > 0,\ k < 0,\ u \in \mathbb{R}$
	$P = \alpha/k$	Generalized Pareto (α,k)	$\alpha > 0,\ k > 0$

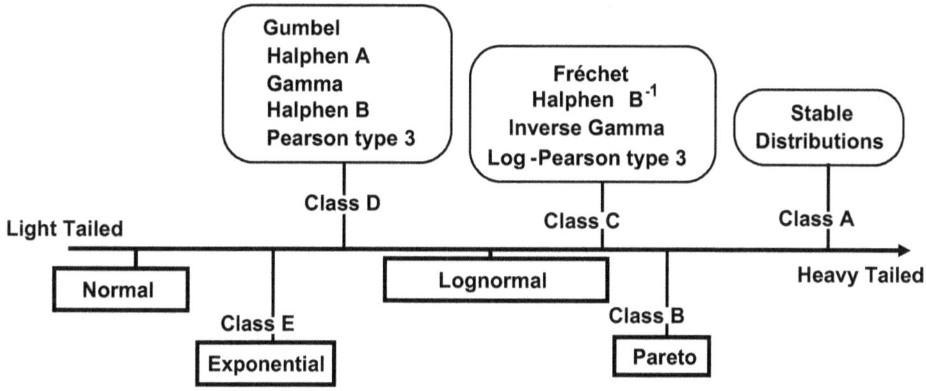

Figure 3.3: Distributions commonly used in hydrology, classified with respect to their tail behavior (El Adlouni et al. 2008).

3.3.2. Asymptotic Behavior for the Halphen Distributions

The Halphen distributions HA and HB belong to the Class D while the HIB has a power tail behavior (Class C). In summary, the Halphen distributions and related distributions have the following asymptotic tail behavior as function of the return period T (Figure 3.4):

- $\ln T$, for the distributions Gamma, Gumbel, and Halphen Type A;
- $\sqrt{\ln T}$, for the Halphen Type B distribution;
- T^{α}, for the Halphen type Inverse B and Inverse Gamma distributions.

Note that almost all distributions considered for risk assessment, in general and especially in hydrological frequency analysis, belong to the Classes C and D or their limiting case (i.e. Lognormal, Figure 3.3). For the Halphen family HA, HB and Gam belong to Class D, and HIB and IG belong to Class C.

In Figure 3.4, the abscise axis corresponds to the return period T and the curves represent the tail behavior of the Halphen distributions and those usually used in frequency analysis, especially in hydrology. Even the distributions of Class C of heavy tailed distribution, for special values of the parameters, may have light tail. These classes are embedded and distributions of the Class C are automatically included in Class D (Figure 3.2). This points out the importance to have (1) optimal approaches for parameter estimation; and (2) efficient criteria to select most adequate model.

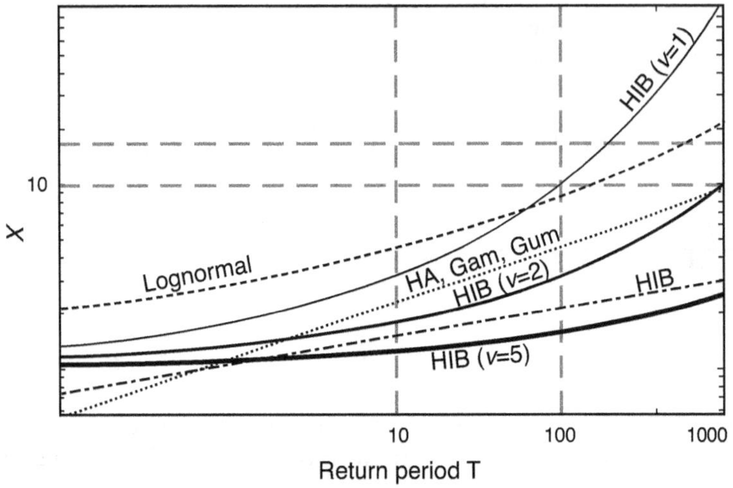

Figure 3.4: Asymptotic behavior of the HA, HB, HIB and some distributions usually used for Frequency Analysis (Lognormal, Gamma, Gumbel).

In order to compare the asymptotic behavior of these distributions, the quantile variation should be considered when the return period T increases. This is equivalent to comparing the derivative functions of the tail behavior of each of the probability distributions. The derivative functions of the tail behavior of the Halphen distributions and related distributions are plotted in Figure 3.5. Note that large values of the derivative function indicate that the tail decreases slowly to zero and thus the distribution is heavy tailed.

Figures 3.4 and 3.5 show that:

- The Gumbel, Gamma, and Halphen Type A distributions allow an identical asymptotic behavior for large values of the random variable. The pdf of the HB, although belonging to the same Class D of sub-exponential distribution, decreases more rapidly than the HA, Gumbel, and Gamma, for large values of the random variable (large return values). For the distributions, Gamma, Gumbel, and Halphen Type A, the behavior is equivalent to $\ln T$; and for Halphen Type B distribution, the behavior is equivalent to $\sqrt{\ln T}$;

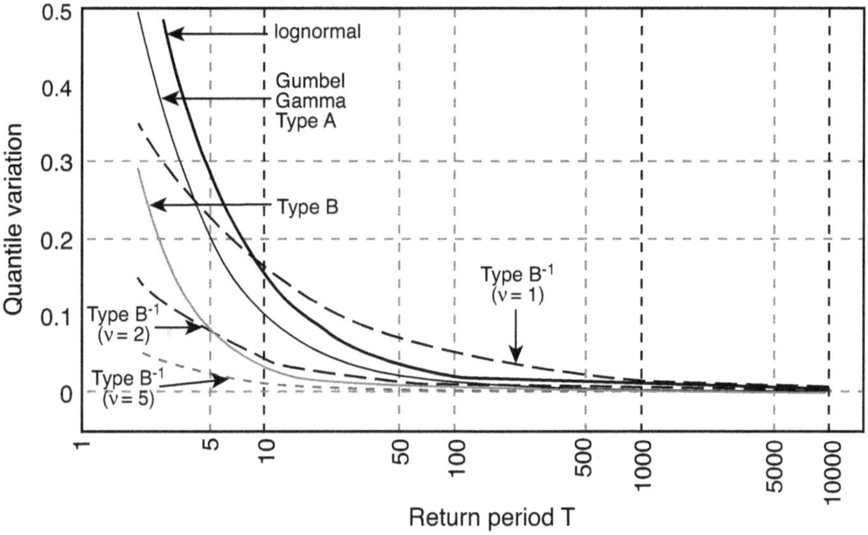

Figure 3.5: Quantile Variation and asymptotic behavior of the Halphen distribution.

- The asymptotic behavior of the Lognormal distribution (Figure 3.4) and its derivative (Figure 3.5) reflect a slower decrease for this fdp than that of the HA, Gumbel, Gamma, HB, and Normal. This type of asymptotic behavior is the basic reason for the development of the Halphen distribution (cf. Chapter 1). Indeed, Halphen noted that the slow decay of the pdf of the Lognormal distribution sometimes proved to be inadequate in empirical studies.
- The asymptotic behavior of the Halphen Type Inverse B (HIB) depends on the shape parameter v. This leads to a large range of decay rates and implies an optimal approach to estimate this shape parameter. For small values of the shape parameter v, the pdf tends very slowly towards zero and the decay rate increases with v (see Figure 3.5).

3.3.3. Decision Support System

A Decision Support System (DSS, Appendix C) was developed to help select the most appropriate class of distributions, with respect to extreme values (El Adlouni and Bobée, 2010). The methods developed in the DSS allow the identification of the most suitable class of distribution to fit a given sample, especially for extremes (Figure 3.3). These methods are adapted to the Halphen family and are illustrated in Figure 3.6.

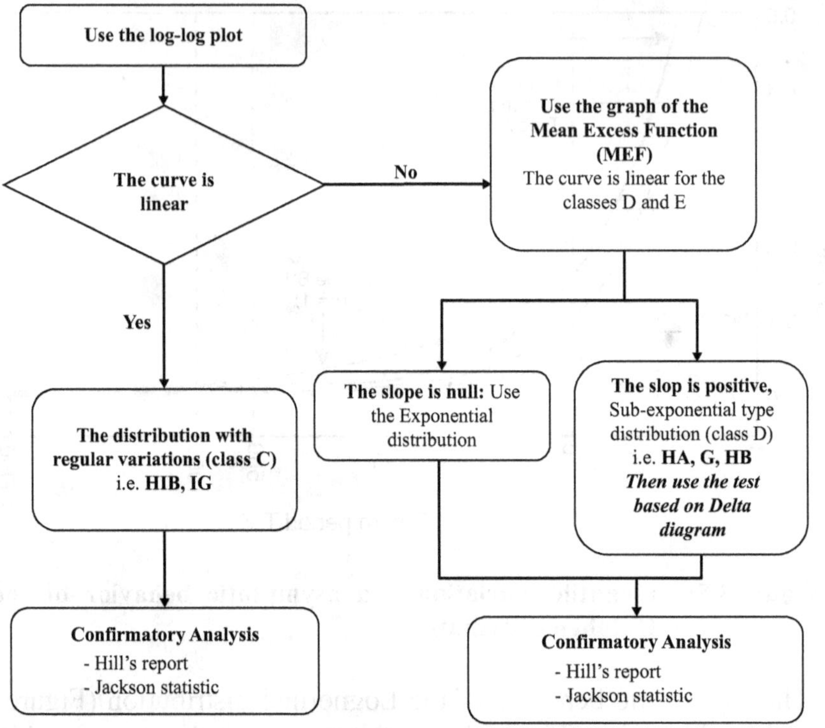

Figure 3.6: Decision Support System (DSS) diagram for Halphen distributions.

The DSS approaches can be summarized by the following steps (Appendix C):

- The Log-Log plot (Appendix C.1): to discriminate between Class C, on one hand; and Classes D and E, on the other hand;
- The mean excess function (MEF, Appendix C.2): to differentiate between Classes D and E; and
- Two statistics: Hill's ratio (Appendix C.3, Hill, 1975) and modified Jackson statistic (Appendix C.4, Jackson, 1967), are used to confirm the analysis of the conclusions suggested by the previous two methods (Appendix C: C.3 and C.4).

More theoretical detail of this classification and the criteria are available in El Adlouni et al. (2008). The DSS approaches are detailed in Appendix C. The corresponding Matlab codes are given in Chapter 6, and with the Halphen Fitting and Generating (HFG) codes and examples.

3.4. THE HALPHEN FAMILY AND THE GENERALIZED EXTREME VALUE (GEV) SYSTEM, A COMPARISON

3.4.1. Introduction

As already mentioned in Chapter 1, several distributions are commonly used in FFA and are generally divided into three groups:

- the normal family (normal, Lognormal),
- the general extreme-value (GEV) family (Gumbel (EV1), Fréchet (EV2), and reverse Weibull (EV3)),
- the Pearson type 3 family (Gamma (Gam), Inverse Gamma (IG), Pearson type 3, Log-Pearson type 3), and
- the Halphen system presented in Chapter 2.

However, the use of the GEV distribution is currently a standard in some countries such as the UK and Australia, due to the simplicity and the explicit form of its quantile function (Eq. (3.6)) and the availability of software for parameter estimation. The GEV distribution appears to be a universal model for hydrological extremes. The theoretical considerations behind the use of the GEV distribution as a standard model, already addressed in Section 3.2, will be discussed and questioned more thoroughly in Section 3.4.2.

As shown in Chapter 2 and Section 3.3, Halphen family of distributions has very attractive mathematical and statistical properties and is suitable for representing a large range of asymptotic behavior.

Almost all distributions given in the last classification belong to the Class D of sub-exponential distributions or the Class C, with regularly varying tail (cf. Section 3.3). Halphen and GEV systems cover both classes. In Figure 3.3, the HA, HB, Gam, and EV1 distributions belong to the Class D (Sub-exponential) while IG, HIB and EV2 belong to the Class C (regularly varying). Both systems, Halphen (HA, HB, and HIB), and GEV (EV1 and EV2), have very interesting mathematical and statistical properties and the objective of the following section is to present the pros and cons of each system.

3.4.2. Theoretical Arguments

Arguments related to the EVT

The use of the GEV distribution in flood frequency analysis can be argued by theoretical results of the EVT. The distribution of the maximum of N random variables, that are independent and identically distributed with parent distribution F_0, converges to one of three possible forms (EV1, EV2, or EV3) of the GEV distribution, when N tends to infinity (Section 3.2). Then, the theoretical properties of the GEV do not justify its systematic use in hydrology (since its theoretical arguments are only apparent, cf. Section 3.2). However, the GEV distributions cover a large type of tails and give more flexibility when fitting extremes and thus should be considered for comparison purposes, as well as other systems, which are not based on theoretical justification (Pearson, Halphen, Lognormal…).

Tail Behavior Argument

Given that the theoretical arguments justifying the systematic use of the GEV distribution and are not fulfilled for hydrological series (cf. Section 3.2), many other distributions are considered, in practice, to fit extremes in hydrology. In addition, Halphen distributions (HA, HB, HIB) with their limiting forms (G and IG, cf. Sections 2.2, 2.3, and 2.4), constitute a complete system (cf. Section 2.5).

In this section, we showed that both families, GEV and Halphen, are flexible models representing different tail behaviors. Although the hydrological datasets do not fulfill EVT assumptions, the GEV distribution remains a good candidate, same as the other distributions for extreme value fitting that have similar tail behavior. However, other considerations and constraints should be taken into account in selecting the most adequate distribution and will be discussed in the following section.

Accuracy and sufficiency of the parameter estimators

In general, for any distribution, under certain conditions:

1) the differentiability of the pdf with respect to the variable and the parameter and;

2) the identifiability of the model (i.e. the pdf is injective as function of the parameter, (Smith, 1985)).

The maximum likelihood (ML) estimators have the desired asymptotic properties (optimal estimation for large sample size). However, these conditions are not fulfilled for the GEV distribution. Indeed, when the shape parameter κ is non-null (Eq. 3.5), which is the case for EV2 and EV3, the support of the random variable depends on the parameters and the pdf of the GEV distribution, but cannot be represented as a member of the exponential family (Eq. (2.23)). This problem generally leads to estimators with very high variance (Smith, 1985). Another problem associated with the ML method, when used for the GEV distribution with small sample sizes, is that the numerical resolution of the ML system can lead to physically impossible estimators of the parameters. To overcome the problems of divergence occurring in the numerical techniques used for ML, Martins and Stedinger (2000) suggest the generalized maximum likelihood approach (GML). In the GML method, a *prior* distributionfor the shape parameter of the GEV model is considered to avoid absurd solutions. Martins and Stedinger (2000) show that the GML estimators are unbiased with low variance.

As shown previously, Halphen system of distributions has very attractive mathematical and statistical properties that are suitable for a large asymptotic behavior (Section 3.2). The Halphen distributions belong to the exponential family of distributions and the ML estimators are optimal, i.e. have minimum variance (Perreault et al. 1999b). These properties will be discussed in more detail in the Chapter 4.

Simplicity

- **Generation procedure:** All the GEV distribution forms are simple and their distribution functions are invertible (the quantile function has an explicit form, (Eq. (3.6)). This is not the case of Halphen distributions. Procedures to generate samples from the GEV distribution are available using the inversion technique and uniform simulated samples. For the Halphen family, the probability densities depend on the Bessel function (HA) or the exponential factorial functions (HB and HIB). Thus the use of the inversion technique for quantile computation or sample generation is not possible based on an explicit form. However, El Adlouni and Bobée (2007) adapted the acceptation-rejection (A-R) approach to generate Halphen variates (cf. Chapter 5). The generation procedures allow to estimate intricate integrals related to these distributions, to determine confidence intervals for quantiles based on the bootstrap approach, or to approximate any probability of exceedance.

- **Parameter estimation:** The well-known parameter estimator procedures (maximum likelihood, method of moments, and L-moments) are available for the GEV distribution and are easy to compute. For Halphen distributions, a numerical approach was proposed by Perreault et al. (1999b) for the maximum likelihood method. However, some numerical approximations are needed to calculate the normalization functions: Bessel and Exponential Factorial. These computational difficulties have not helped the widespread use of these distributions for FFA.

This problem can now be overcome by using Matlab codes (HFG, Chapter 6) considering the method of moments (MM) developed for

- Halphen type A distribution (Fitzgerald, 2000), and
- Halphen type B and Inverse B (Chebana et al. 2008).

More detail on the Halphen parameter estimation methods are given in Chapter 4.

The HFG Matlab codes also contain a Decision Support System (DSS, cf. Chapter 6) to discriminate between the Classes C, D, and E. The methods developed in DSS allow the identification of the most adequate class of distribution to fit a given sample, especially for extremes. These methods are introduced in Section 3.3.3 and are detailed in Appendix C.

The Matlab codes corresponding to the DSS, to identify the class of the tail behavior, are available and their use is described in Chapter 6.

3.5. TESTING HA AND HB AGAINST THE GAMMA DISTRIBUTION

The study of the tail behavior of the Halphen family distributions, shows that the HA and HB distributions belong both to the Class D of sub-exponential distributions, while, the HIB distribution has a regularly varying tail (Class C) (El Adlouni et al., 2008). Thus, the tail behavior approach, based on the DSS, makes it is possible to differentiate between HA and HB distributions on one hand, and HIB on the other hand. However, for a given sample, this approach doesn't allow a choice between the HA, HB, and Gamma distributions, since they belong to the same Class D (sub-exponential distributions).

The main objective of the present section is to present the extension of the Decision Support System (DSS), developed to discriminate between HA and HB and their limiting distribution Gamma (Figure 2.1). The developed test is summarized here, based on the diagram (δ_1, δ_2). For more detail see El Adlouni et al. (2015).

3.5.1. Hypothesis and Alternatives

The main objective of the developed test is to discriminate between the HA, HB, and the Gamma distributions presented in the (δ_1,δ_2) diagram (Figure 2.1). The test is included in the Decision Support System (DSS) of the HFG software (Chapter 6) for choosing the most appropriate fit, when Class D is selected. HA, HB, and Gamma distributions have a sub-exponential tail behavior (Class D), and the ability of such a test to differentiate between these distributions will complete the DSS (El Adlouni and Bobée, 2010). Thus, the test developed will be used for a given sample of Independent and Identically Distributed (IID) dataset $(x_1,\cdots,x_n)$ with size *n* for the following hypotheses:

- H_0: The variable follows the Gamma distribution (null hypothesis),
- H_1: The variable follows the HA or the HB distribution (alternative hypotheses).

3.5.2. Moment Ratio Diagram: Test Elaboration

The objective of the proposed approach is to test the null hypotheses H_0, (the variable following the Gamma distribution), against the alternative hypotheses H_1, (the variable following the HA or the HB distribution). It is built based on the determination of acceptance/rejection regions with a significance level α_0 (generally α_0=5% or 1%), in the (δ_1,δ_2) moment ratios' diagram. The test is developed for a significance level $\alpha_0 = 0.05$. Thus the decision rule will be based on the estimated moment ratios (δ_1,δ_2)-diagram, using the observed data. Note that the corresponding theoretical moments in the case of the Gamma distribution are given by: $A=\lambda$, $\ln G=\psi(\lambda)$, and $H=\lambda-1$; where $\psi(\lambda)$ is the Digamma function defined in Bobée and Ashkar (1991) (Appendix A). The Gamma curve, in the moment ratios (δ_1,δ_2)-diagram, depends only on the shape parameter λ (El Adlouni et al. 2015), and δ_1 and δ_2 are given by:

$$\delta_1 = \ln\frac{A}{G} = \ln\lambda - \psi(\lambda) \tag{3.10}$$

$$\delta_2 = \ln\frac{G}{H} = \psi(\lambda) - \ln(\lambda - 1) \tag{3.11}$$

The derivative of the Gamma curve in the (δ_1,δ_2)-diagram is given by:

$$y' = \frac{d\delta_2}{d\delta_1} = \frac{d\delta_2}{d\lambda}\cdot\frac{1}{\frac{d\delta_1}{d\lambda}}, \quad \text{and then} \quad y' = \frac{d\delta_2}{d\delta_1} = \frac{\psi'(\lambda) - \left(\frac{1}{\lambda-1}\right)}{\left(\frac{1}{\lambda}\right) - \psi'(\lambda)}$$

Where $\psi'(\lambda)$ is the Trigamma function (Bobée and Ashkar, 1991, Appendix A). This can be used to approximate the curve using first order Taylor series.

Note that for the Gamma distribution the coefficient of skewness is related to the shape parameter through:

$$C_s = \frac{2}{\sqrt{\lambda}} \quad \text{or equivalently} \quad \lambda = \frac{4}{C_s^2} \tag{3.12}$$

Then the shape parameter can be estimated from the coefficient of skewness.

For a given sample $x_1,\ldots,x_n$, the estimated moment ratio $(d_{1,n},d_{2,n})$ of the theoretical moments (δ_1,δ_2), are deduced from Eqs. (3.13) and (3.14), where the moments A, G, and H are replaced by their sample estimations $\hat{A}$, $\hat{H}$, and $\hat{G}$:

$$\begin{aligned} d_{1,n} &= \ln\hat{A} - \ln\hat{G} \\ &= \ln\left(\frac{1}{n}\sum_{i=1}^{n} x_i\right) - \frac{1}{n}\sum_{i=1}^{n}\ln x_i \end{aligned} \tag{3.13}$$

$$\begin{aligned} d_{2,n} &= \ln\hat{G} + \ln\hat{H}^{-1} \\ &= \frac{1}{n}\sum_{i=1}^{n}\ln x_i + \ln\left(\frac{1}{n}\sum_{i=1}^{n}\frac{1}{x_i}\right) \end{aligned} \tag{3.14}$$

where

$$\hat{A}_j = \frac{1}{n}\sum_{i=1}^{n} x_i, \quad \ln\hat{G}_j = \left(\frac{1}{n}\sum_{i=1}^{n}\ln x_i\right) \quad \text{and} \quad \hat{H}_j^{-1} = \frac{1}{n}\sum_{i=1}^{n}\frac{1}{x_i}$$

are, respectively, the estimations of the arithmetic, geometric, and harmonic means (Eqs. (2.24a-c)). The determination of the acceptance region of H_0, for a significance level α_0, is performed using the following steps:

Algorithm 1

1. *Selection of a particular value of λ ranged from 1.108 and 400 (corresponding to C_s values, between 0.1 to 1.9 (Eq. (3.12)));*

2. *Calculation of theoretical values of δ_1 and δ_2 for fixed λ Eqs. (3.10) and (3.11);*

3. *Generation of N samples of size n for the chosen value λ, in (step 1), from the Gamma distribution $G1(l)$, (cf. Bobée and Ashkar, 1991, Appendix F);*

4. *Estimation $d_{1,n}$ and $d_{2,n}$, of the moment ratios δ_1 and δ_2, for each sample (Eqs. (3.13) and (3.14)). A scatter plot is then obtained, centered in the point (δ_1, δ_2), corresponding to the theoretical value λ, already fixed;*

5. *Computation of the Euclidean distance between the different points representing the generated samples and the fixed one on Gamma curve, corresponding to the theoretical λ value already fixed. The expression of the distance is given by:*

$$dist = \left(\left(\delta_1 - d_{1,n} \right)^2 + \left(\delta_2 - d_{2,n} \right)^2 \right)^{1/2} \tag{3.15}$$

   δ_1 and δ_2, already given in step 2;

6. *Determination for each value of the sample size n, of the acceptance zone (Accept H_0), containing $(1-\alpha_0)$% of the generated samples corresponding to the $(1-\alpha_0)$% of the nearest points to the theoretical point (δ_1, δ_2), according to the Euclidean distance (Eq. (3.15)).*

When applying the proposed approach, we obtain an envelope region of the Gamma curve containing 95% of the samples (case where $\alpha_0 = 5\%$). Thus, the generated samples around each fixed point on the Gamma curve are divided into two parts according to its relative position with respect to the curve (above or below).

In step 6, the identified points forming the confidence zones are the farthest points to the Gamma curve which are perpendicular to the curve at the vicinity of the theoretical point (δ_1,δ_2), corresponding to the already fixed λ value. The equation of the perpendicular to the curve is given by $y=(-(x-x_0)/y_0')+y_0$; where (x_0,y_0) are the coordinates of the fixed points on the Gamma curve, and y_0' is the value of the derivative of the point of δ_2 and a function of δ_1, evaluated at the same points previously set on the Gamma curve.

According to Eq. (3.12), an exact expression links the shape parameter λ and the coefficient of skewness C_s. Thus the values of the shape parameter, considered in this study, correspond to a coefficient of skewness between 0.1 and 2.5, (with λ varying between 0.64 and 400) which corresponds to the range generally encountered in hydrology. The identified samples, corresponding to this significance level, are considered to define the envelope curves for the Gamma distribution, as illustrated in Figure 3.7.

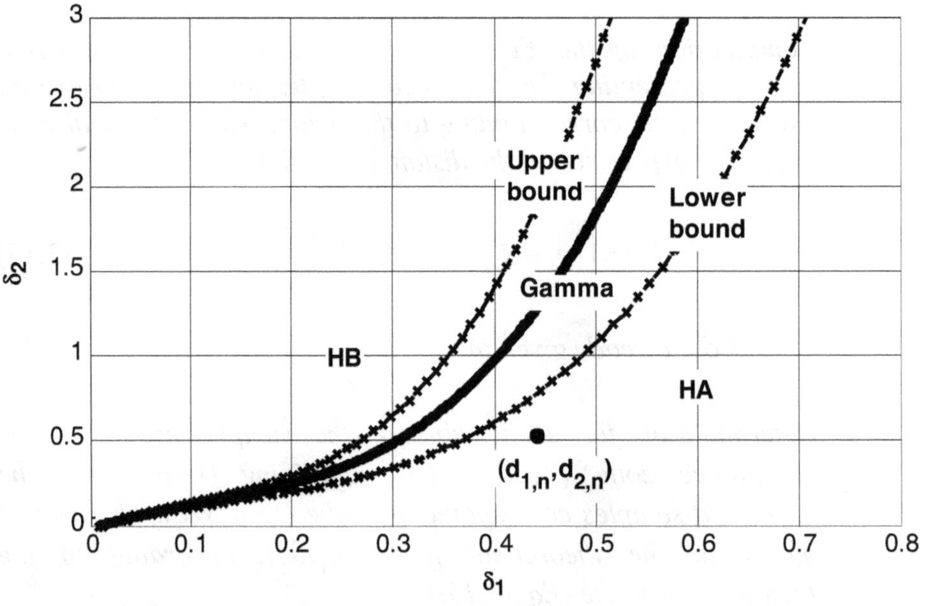

Figure 3.7: Confidence zones around the Gamma curve, illustration for n=100.

Several fixed sample sizes (*n* = *30, 50, 80, 100,* and *200*) have been considered to develop the critical regions of the test. Thus for a sample

$x_1,\ldots,x_n$, the two corresponding empirical moment ratios $(d_{1,n},d_{2,n})$. Equations (3.13) and (3.14) are represented in the (δ_1,δ_2) -diagram (Figure 2.1) in order to decide on the adequacy of the Gamma distribution through its position with respect to the envelope curves. Note that the empirical moments are always biased, especially for samples with finite sizes, so it's important to correct this bias (Bobée and Robitaille, 1975). In the proposed test, the bias has been corrected, especially for δ_2 ; the bias of δ_1 is not significant (El Adlouni et al. 2015). Then, in the Algorithm 1 (cf Section 3.5.2) the empirical moment ratio δ_2 is considered, with corrected bias. More detail on the test formulation and simulation results can be found in El Adlouni et al. (2015).

3.5.3. Power Study

The power study shows (El Adlouni et. al., 2015) that for almost all shapes of the HA and HB distributions, the test allows to differentiate between the Gamma distribution and both alternatives. Note that when the alternative corresponds to the HA distribution, the power of the test increases with the coefficient of skewness. However, in the case of the HB alternative, the power of the test decreases when the coefficient of skewness increases. This is mainly due to the position of the theoretical distribution in the delta diagram.

For both alternatives, the power increases with the sample size between 30 and 200. For the HA case, the power of the test is an increasing function of the skewness coefficient and varies from 71% (moderate skewness $C_s = 1.2$) to 88% (high skewness $C_s = 2.5$) in the case of sample size 30. The power in the HA cases is larger than 86%, for all skewness coefficients, when the sample size is equal to 100. For the HB cases, the power of the test is a decreasing function of the skewness coefficient and is varying from 33% (skewness around $C_s = 0.77$) to 93% (low skewness $C_s = 0.28$), in the case of sample size 100. Note that even for cases with low power test, this diagram allows for differentiation between the HA and HB distributions. All simulated samples for HA and HB distributions, belong to the region of the original distribution in the (δ_1,δ_2)-diagram (El Adlouni et al. 2015).

3.5.4. Conclusion

The proposed test allows for differentation between the HA, HB, and the Gamma distributions, which belong all three to the Class D of sub-exponential distributions. The methodology involves:

- the bias correction of the estimated moment ratio δ_2,
- the determination of the critical region corresponding to the null hypothesis (H_0: The sample follows a Gamma distribution) and,
- a power study based on alternatives with usually observed coefficient of skewness.

Results show that for almost all cases, and even for small sample sizes (n=30); the power of the test is acceptable especially for the HA alternatives. For the HB cases, the power of the test is high especially for low coefficient of skewness. The proposed test represents a substantial improvement to complete the Decision Support System (DSS, Figure 3.6) currently available in HFG software (available with the present monograph, Chapter 6). Presently DSS allows to differentate between distributions of the Classes C, with regularly varying tail, and Class D of sub-exponential tail. Given that the HA, HB, and Gamma distributions belong to the same Class D, the proposed test represents a useful tool to discriminate between them.

Chapter 4
Parameter Estimation Methods for the Halphen Distributions

4.1. INTRODUCTION

To estimate the parameters of each Halphen distribution, the maximum likelihood (ML) method was proposed by Perreault el al. (1999b). The ML estimation is optimal for Halphen distributions since it is based on three statistics jointly sufficient (cf. Chapter 2). However, the resolution of the ML system is complex because it requires some manipulations of the derivatives of the normalization functions (Bessel and exponential factorial). To solve the ML system efficiently, a numerical method based on profile likelihood was presented by Perreault et al. (1999b). It corresponds to numerical approximations which do not lead to an exact solution. Furthermore, the bias of the ML is asymptotically null, but it may be large for small sample sizes with a large root-mean square-error. Consequently, despite the sufficiency of ML estimators, it is relevant to consider other estimation methods for small sample sizes encountered in hydrology. Fitzgerald (2000) proposed a method of moments (MM) to estimate the parameters of the HA distribution. An advantage of the method of moments is the explicit forms of estimators with respect to the moments, thus facilitating its implementation and making computations faster. As a natural extension of the work of Fitzgerald (2000), Chebana et al. (2008) developed a method of moments to estimate the parameters of HB and HIB distributions.

To obtain the ML estimators for the three Halphen distributions, Perreault et al. (1999b) carried out a partial maximization of the conditional likelihood function $L(\nu \mid \alpha(\nu), m((\nu))$, for a fixed value of the parameter u. In addition, Chebana et al. (2010) propose two new mixed methods that combine the advantages of ML (optimality) and of MM (fast computations):

- The first method MMD is direct where the shape parameter ν is estimated using MM equation, which leads to a value of ν, close to the optimum value, and then estimators of α and m are obtained from the two corresponding equations of the ML system.

- The second method (Iterative method of moments; IMM) is iterative. It consists in varying ν from its estimator obtained by MM, and stop at the value of ν which corresponds to maximizing the conditional log-likelihood function $\log L(\nu \mid \alpha(\nu), m(\nu))$, and then α and m are estimated from the two corresponding equations of ML.

On the other hand, note that the performances of the MM method for HA distribution have not been evaluated and have not been compared with ML. However, Chebana et al. (2008) have made a comparison between the estimates by MM and ML for HB and HIB distributions. Thus in this chapter, we present the main estimation methods to estimate the parameters of the three Halphen distributions along with comparative study.

4.2. MAXIMUM LIKELIHOOD METHOD (ML)

4.2.1. ML for the HA Distribution

Let $\underline{X} = (X_1, X_2, \ldots, X_n)$ be a sample of *n*, independent and identically distributed random variables drawn from the HA distribution and $\underline{x} = (x_1, x_2, \ldots, x_n)$, a realization of $\underline{X}$. The maximum likelihood estimators (MLE), of the three parameters, are generally deduced by numerical maximization of the likelihood function or its logarithm.

In general, when the distribution belongs to the exponential family (as in the case for the three Halphen distributions), the MLE equations are related to the sufficient statistics. We recall here the general pdf form, of the exponential family with the vector of the parameters $\underline{\theta} = (\theta_1, \theta_2, \theta_3)$, is given by Eq. (2.23)

$$\left(f(x) = \exp\left\{ \sum_{i=1}^{3} c_i(\theta_1, \theta_2, \theta_3) T_i(x) + d(\theta_1, \theta_2, \theta_3) + S(x) \right\} \right).$$

The likelihood function is given by:

$$L(\theta; x_1, \ldots, x_n) = \prod_{j=1}^{n} f(x_j; \theta) = \exp\left\{ \begin{aligned} &\left(\sum_{j=1}^{n} \sum_{i=1}^{3} c_i(\theta_1, \theta_2, \theta_3) T_i(x_j) \right) + \\ &\left(n d(\theta_1, \theta_2, \theta_3) + \sum_{j=1}^{n} S(x_j) \right) \end{aligned} \right\} \quad (4.1)$$

where $T(\underline{x})=(T_1(\underline{x}),T_2(\underline{x}),T_3(\underline{x}))$ are sufficient statistics for the probability density function of $\underline{X}$.

To determine the MLE vector $\underline{\theta}_{ML}$ of the parameters, the likelihood function, L, should be maximized. This is equivalent to the maximization of the logarithm of the likelihood function $\ln L$, The log-likelihood function in the case of an exponential family distribution is given by:

$$\ln L(\theta;x_1,\ldots,x_n)=\sum_{j=1}^{n}\sum_{i=1}^{3}c_i(\theta_1,\theta_2,\theta_3)T_i(x_j)+nd(\theta_1,\theta_2,\theta_3)+\sum_{j=1}^{n}S(x_j).$$

Then,

$$\ln L(\theta;x_1,\mathrm{K}\ ,x_n)=\left(\sum_{j=1}^{n}\sum_{i=1}^{3}c_i(\theta_1,\theta_2,\theta_3)T_i(x_j)\right)+\left(nd(\theta_1,\theta_2,\theta_3)+\sum_{j=1}^{n}S(x_j)\right) \tag{4.2}$$

As in Section 2.2.3 (Eqs. (2.24a,b,c)), it can be shown that for a given sample $\underline{x}=(x_1,x_2,\ldots,x_n)$ from HA distribution, the following statistics are sufficient

$$T_1(\underline{x})=\frac{1}{n}\sum_{j=1}^{n}T_1(x_j);\ T_2(\underline{x})=\frac{1}{n}\sum_{j=1}^{n}T_2(x_j)\ \text{and}\ T_3(\underline{x})=\frac{1}{n}\sum_{j=1}^{n}T_3(x_j)$$

and thus, the maximum likelihood system is given by:

$$E[T_1(X)]=\frac{1}{n}\sum_{j=1}^{n}T_1(x_j) \tag{4.3a}$$

$$E[T_2(X)]=\frac{1}{n}\sum_{j=1}^{n}T_2(x_j) \tag{4.3b}$$

$$E[T_3(X)]=\frac{1}{n}\sum_{j=1}^{n}T_3(x_j) \tag{4.3c}$$

where $E\left[T_j(X)\right]_{j=1,2,3}$ depends on the parameters vector $\underline{\theta}$.

If the system admits a unique solution for the parameter vector $\theta=(\theta_1,\theta_2,\theta_3)$, then it will correspond to the maximum likelihood estimators of θ (Bickel et Doksum, 1977). The system of Eq. (4.3) is equivalent to the generalized method of moment (Bobée and Ashkar, 1988) where the moments are replaced by the sufficient statistics (A, G, and H). The importance of the sufficient statistics has been outlined in Chapter 2. The existence of such statistics optimally ensures the MLE. Therefore, to develop the MLE of the Halphen type A parameters, the expectation of the sufficient statistics

$$T_1(\underline{x})=\frac{1}{n}\sum_{i=1}^{n}\ln x_i \;;\; T_2(\underline{x})=\frac{1}{n}\sum_{i=1}^{n}x_i \text{ and } T_3(\underline{x})=\frac{1}{n}\sum_{i=1}^{n}\frac{1}{x_i}$$

are considered. The corresponding empirical moments are, respectively, the logarithm of the geometric mean (Log G), the arithmetic (A), and harmonic (H) means. The ML system of the Halphen type A distribution can then be developed using Eqs. (2.11), (2.12), and (2.13) corresponding to the expectation of the sufficient statistics of the HA, and their corresponding empirical counterpart. The system for the HA distribution is then given by:

$$m\frac{K_{v+1}(2\alpha)}{K_v(2\alpha)}=\frac{1}{n}\sum_{i=1}^{n}x_i=A \tag{4.4a}$$

$$\frac{1}{m}\frac{K_{v-1}(2\alpha)}{K_v(2\alpha)}=\frac{1}{n}\sum_{i=1}^{n}\frac{1}{x_i}=H^{-1} \tag{4.4b}$$

$$\ln m+\frac{\partial K_v(2\alpha)/\partial v}{K_v(2\alpha)}=\frac{1}{n}\sum_{i=1}^{n}\ln x_i=\ln G \tag{4.4c}$$

Solving the last system of equations requires evaluating the Bessel function and its first derivative, which does not have an explicit form. The complexity of the resolution of the ML equation system, in practice, has limited their use. For this reason, to make the Halphen system easier to use, the numerical methods presented in this section are available in HFG software (Chapter 6).

To solve this system in terms of the parameters of the HA distribution, Perreault et al. (1999b) proceeded the same as Jørgensen (1982) did for the Generalized Inverse Gaussian (GIG, cf. Section 2.2.4) distribution. The estimation process is done in two stages. First, the parameters α and m are estimated for a range of fixed values of v. Note their estimators $\hat{\alpha}$ and $\hat{m}$, respectively. Then, the partially maximized log-likelihood function, $\ln L_{HA}\left(v|\hat{\alpha},\hat{m}\right)$, is used to obtain an estimate $\hat{v}$ of the parameter v. When maximum likelihood estimation of (α,m) for fixed v is considered, only two equations are needed. By arranging the first two equations in Eq. (4.4) and taking into account the previous Eqs. (2.17) and (2.18), we get the following equation:

$$D_A(\alpha,v)=\frac{K_{v+1}(2\alpha)K_{v-1}(2\alpha)}{\left[K_v(2\alpha)\right]^2} \quad \text{and} \quad R_A(\alpha,v)=\frac{K_{v+1}(2\alpha)}{K_v(2\alpha)}$$

Then the following system is obtained:

$$D_A\left(\alpha,v\right)=\frac{A}{H} \tag{4.5a}$$

$$m=\frac{A}{R_A\left(\alpha,\ v\right)} \tag{4.5b}$$

Provided that a solution exists for a given value of v; $\hat{\alpha}$ may first be obtained by solving Eq. (4.5a) numerically, and then $\hat{m}$ may be obtained directly from Eq. (4.5b). The properties of the function $D_A(\alpha,v)$ are important for the estimation of the HA parameters and are presented in more detail in the Appendix A.

Let $U=AH^{-1}/[AH^{-1}-1]$, ($U>1$ since $H\leq A$), then (Perreault et al, 1999b):

a) If $v\geq U$ then the limiting distribution, Gamma is a much better fit the studied sample; and;

b) if $v\leq -U$, then the limiting distribution, Inverse Gamma should be considered rather than the HA (Figure 2.4).

c) the ML estimators of the parameters α and m are solution of the system given by Eqs. (4.5a,b) if, and only if, the parameter v belongs to the interval $\left]-U;U\right[$.

When the appropriate interval $\left]-U;U\right[$, for the shape parameter v, is defined, the derivative of the partial likelihood function is evaluated on the bounds (Perreault et al. 1999b). Their expressions are, respectively, given in Eqs. (4.6a) and (4.6b), respectively, such that:

$$l'_{HA}(-U) = n\left[\ln\left(\frac{G}{H}\frac{1}{U}\right) + \Psi(U)\right] \tag{4.6a}$$

$$l'_{HA}(U) = n\left[\ln\left(\frac{G}{A}U\right) - \Psi(U)\right] \tag{4.6b}$$

Where $\Psi(z) = \partial[\ln\Gamma(z)]/\partial z$ is the digamma function (Bobée and Ashkar, 1991, Appendix A).

Thus, the ML procedure can be summarized by the following steps:

a) If $l'_{HA}(U) > 0$ and $l'_{HA}(-U) > 0$, then the value of n that maximizes the log-likelihood $\ln L_{HA}(v\,|\,\alpha,m)$, is greater than U, and the Gamma distribution will be more suitable to fit the sample than HA;

$$\ln L(v\,|\alpha,m) = n\left[v\ln\left(v/A\right) - \ln\Gamma\left(v\right) + \left(v-1\right)\ln G - v\right] \tag{4.7a}$$

b) If $l'_{HA}(U) < 0$ and $l'_{HA}(-U) < 0$, then the value of v that maximizes $\ln L_{HA}$, is smaller than $-U$, and the Inverse Gamma distribution will be more suitable to fit the sample than HA; and the log-likelihood function is given by:

$$\ln L(v\,|\alpha,m) = n\left[-v\ln\left(-vH\right) - \ln\Gamma\left(-v\right) + \left(v-1\right)\ln G + v\right] \tag{4.7b}$$

c) If $l'_{HA}(U) < 0$ and $l'_{HA}(-U) > 0$, then the value of v that maximizes $\ln L_{HA}$ belongs to the interval $]-U;U[$; and the ML estimators are solutions of the system of Eqs. (4.5a,b). In this case, the partial log-likelihood function is given by (Perreault et al. 1999b):

$$\ln L_{HA}(v\,|\,\alpha,m) = n\left\{\ln\left[\frac{G^{v-1}}{2m^{v}K_{v}(2\alpha)}\right] - \frac{\alpha\left[K_{v+1}(2\alpha) + K_{v-1}(2\alpha)\right]}{K_{v}(2\alpha)}\right\} \tag{4.7c}$$

These steps are considered to maximize the profile log-likelihood through a discrete partition of the interval $]-U;U[$, to obtain the optimal value of $\hat{\nu}$ and then the corresponding values $\hat{\alpha}$ and $\hat{m}$, using (Eqs. 4.5).

4.2.2. ML for the HB Distribution

The maximum likelihood system for the Halphen type B distribution can be deduced from Eqs. (2.32 -2.34) and is given by:

$$m\frac{ef_{\nu+\frac{1}{2}}(\alpha)}{ef_{\nu}(\alpha)}=\frac{1}{n}\sum_{i=1}^{n}x_i=A \tag{4.8a}$$

$$m^2\frac{ef_{\nu+1}(\alpha)}{ef_{\nu}(\alpha)}=\frac{1}{n}\sum_{i=1}^{n}x_i^2=Q \tag{4.8b}$$

$$\ln m+\frac{\partial ef_{\nu}(\alpha)/\partial\nu}{2ef_{\nu}(\alpha)}=\frac{1}{n}\sum_{i=1}^{n}\ln x_i=\ln G \tag{4.8c}$$

where A, Q, and G are the arithmetic, quadratic, and geometric means (Eqs 2.42a,b,c), respectively. These moments represent the sufficient statistics of the HB distribution (cf. Section 2.3.3).

Similarly to the ML procedure for the HA distribution, then Eq. (4.8) can be solved by numerical approach. The ML solving steps remain the same.

For fixed ν, $\alpha(\nu)$, and $m(\nu)$ are solutions from Eqs. (4.8a,b), which can be rewritten as:

$$D_B(\alpha(\nu),\nu)=\frac{Q}{A^2} \tag{4.9a}$$

$$m(\nu)=\frac{A}{R_B(\alpha(\nu),\ \nu)} \tag{4.9b}$$

Taking into account Eqs. (2.37) and (2.38) recalled below

$$R_B(\alpha,\nu)=\frac{ef_{\nu+1/2}(\alpha)}{ef_{\nu}(\alpha)}\quad\text{and}\quad D_B(\alpha,\nu)=\frac{ef_{\nu+1}(\alpha)ef_{\nu}(\alpha)}{\left[ef_{\nu+1/2}(\alpha)\right]^2}.$$

Let $V=1/[2(Q/A^2-1)]$, the ML estimators of the parameters α and m are solutions from Eqs. (4.9a,b) if, and only if, ν is in the interval $]0;V[$ (Perreault et al. 1999b).

a) If $v \geq V$, then the limiting distribution Gamma, will be preferable to fit the sample (cf. Figure 2.5).

b) If and only if v belongs to the interval $]0;V[$, the derivative of the partial log-likelihood function $\ln L_{HB}$, is given by the equation:

$$l'_{HB}(V) = 2n\left[\ln\left(\frac{G}{A}2V\right) - \Psi(2V)\right] \tag{4.10}$$

Then the ML procedure to estimate the parameters of the HB distribution can be summarized by the following steps:

a) If $l'_{HB}(V) \geq 0$, the maximum of partial log-likelihood function $\ln L_{HB}$, with respect to v, is greater than the upper bound V. In this case, the limiting distribution Gamma, will be a preferable fit to the sample than HB;

b) If $l'_{HB}(V) < 0$, the maximum of $\ln L_{HB}$ belongs to the interval $]0;V[$, and then the partial log-likelihood function is given by (Perreault et al. 1999b):

$$\ln L_{HB}(v \mid \alpha, m) = n\left\{\ln\left[\frac{2G^{2v-1}}{m^{2v} ef_v(\alpha)}\right] - \frac{ef_{v+1}(\alpha)}{ef_v(\alpha)} + \hat{\alpha}\frac{ef_{v+1/2}(\alpha)}{ef_v(\alpha)}\right\} \tag{4.11}$$

These steps are considered to maximize the profile log-likelihood through a discrete partition of the interval $]0;V[$ to obtain the optimal value $\hat{V}$, and then the corresponding $\hat{\alpha}$ and $\hat{m}$ (Eqs. 4.9).

4.2.3. ML for the HIB Distribution

As shown in Section 2.4.1, the HIB distribution is related to the HB distribution through symmetric relationship (Eq 2.44). Then, the ML system for the HIB distribution can be deduced from that of the HB by substituting m by m^{-1} and x by x^{-1} in the system of Eqs. (4.8a,b,c) and is given by:

$$\frac{1}{m}\frac{ef_{v+(1/2)}(\alpha)}{ef_v(\alpha)} = \frac{1}{n}\sum_{i=1}^{n}\frac{1}{x_i} = \frac{1}{H} \tag{4.12a}$$

$$\frac{1}{m^2}\frac{ef_{\nu+1}(\alpha)}{ef_\nu(\alpha)}=\frac{1}{n}\sum_{i=1}^{n}\frac{1}{x_i^2}=\frac{1}{IQ} \tag{4.12b}$$

$$\ln m-\frac{\partial ef_\nu(\alpha)/\partial\nu}{2ef_\nu(\alpha)}=\frac{1}{n}\sum_{i=1}^{n}\ln x_i=\ln G \tag{4.12c}$$

where *H, IQ,* and *G* are, respectively, the Harmonic, Inverse Quadratic, and Geometric means. As for the HA and HB distributions, this system is based on sufficient statistics of the HIB distribution (cf. Section 2.4.3).

To solve this system, a similar procedure used for HA and HB could be considered for the HIB system. Thus the three steps shown for the HA and HB remains the same.

For fixed value of ν, $\alpha(\nu)$, and $m(\nu)$ are solutions of the system of Eqs. (4.12a,b), which is equivalent to:

$$D_B(\alpha(\nu),\nu)=\frac{H^2}{IQ} \tag{4.13a}$$

$$m(\nu)=H.R_B(\alpha(\nu),\nu) \tag{4.13b}$$

where the functions D_B and R_B are given, respectively, by Eqs. (2.37) and (2.38).

Let $W=1/[2(H^2IQ^{-1}-1)]$ (Perreault et al. 1999b),

a) If $\nu\geq W$, then the Inverse Gamma distribution will be preferable to represent the sample (Figure 2.6).

b) if, and only if, ν belongs to the interval $]0;W[$, then the ML estimators of the parameters α and m are solutions of Eqs. (4.13a,b).

In order to compare the estimated value of ν to the bound W, the value of the derivative of the partial likelihood function $l'_{HIB}(W)$:

$$l'_{HIB}(W)=2n\left[\ln\left(\frac{H}{G}2W\right)-\Psi(2W)\right] \tag{4.14}$$

Then the following steps are considered to compute the ML estimators of the three parameters, when the solution exists.

a) If $l'_{HIB}(W) \geq 0$, the maximum of the log-likelihood $\ln L_{HIB}$, with respect to v, is greater than W. In this case, the Inverse Gamma distribution is more appropriate to represent the sample; and the likelihood is given by Eq. (4.7b):

$$\ln L(\omega, v) = n\left[2v\ln\omega - \ln\Gamma(2v) - (2v+1)\ln G - \omega H^{-1}\right]$$

b) If $l'_{HIB}(W) < 0$, then the maximum of the function $\ln L_{HIB}$, with respect to v, belongs to the interval $]0;W[$. In this case, the partial log-likelihood function is given by (Perreault et al. 1999b):

$$\ln L_{HIB}(v \mid \alpha, m) = n\left\{\ln\left[\frac{2m^{2v}}{G^{2v+1} ef_v(\alpha)}\right] - \frac{ef_{v+1}(\alpha)}{ef_v(\alpha)} + \alpha\frac{ef_{v+1/2}(\alpha)}{ef_v(\alpha)}\right\} \quad (4.15)$$

These steps are considered to maximize the profile log-likelihood through a discrete partition of the interval $]0;W[$, to obtain the optimal value $\hat{v}$ using (Eq. (4.15)), and then, the corresponding values $\hat{\alpha}$ and $\hat{m}$ (Eq. (4.13)).

4.3. THE METHOD OF MOMENT (MM)

4.3.1. MM for HA Distribution

For the HA distribution, the Method of Moment (MM) method, proposed by Fitzgerald (2000), is based on, (1) the differential equation of Bessel function and its recurring formula (Abramovitz and Stegun, 1972):

$$\alpha^2\frac{\partial^2 K_v(\alpha)}{\partial\alpha^2} + \alpha\frac{\partial K_v(\alpha)}{\partial\alpha} - (\alpha^2 + v^2) = 0 \quad (4.16)$$

$$\frac{1}{\alpha}\frac{\partial}{\partial\alpha}\left(\alpha^v K_v(\alpha)\right) = -\alpha^{v-1}K_{v-1}(\alpha) \quad (4.17)$$

and, (2) the expression of the moments for any order r given by Eq. (2.10). For the three parameters, Fitzgerald (2000) obtains the following explicit estimators:

$$m^2 = \frac{E\left(X^{-1}\right)\mathrm{Var}(X) - E(X)\left(E(X)E\left(X^{-1}\right) - 1\right)}{E\left(X\right)\mathrm{Var}\left(X^{-1}\right) - E\left(X^{-1}\right)\left(E(X)E\left(X^{-1}\right) - 1\right)} \quad (4.18)$$

$$\alpha = \frac{m^{-1}E(X) - mE\left(X^{-1}\right)}{m^{-2}\mathrm{Var}(X) - m^{2}\mathrm{Var}\left(X^{-1}\right)} \tag{4.19}$$

$$\nu = \frac{\left(m^{-1}E(X) - mE\left(X^{-1}\right)\right)^{2}}{m^{-2}\mathrm{Var}(X) - m^{2}\mathrm{Var}\left(X^{-1}\right)} \tag{4.20}$$

In Eqs. (4.18), (4.19), and (4.20), the theoretical moments are generally unknown. In order to obtain the estimators, those moments should be replaced by their estimates from the sample $x_1,\ldots,x_n$ of size *n*, i.e.,

$$E(X^r) \text{ is replaced by } n^{-1}\sum_{i=1}^{n} x_i^r \,.$$

4.3.2. MM for HB Distribution

As shown in Section 2.3.2, the rth non-central moment of a random variable *X*, following $f_{HB}(y;\nu,m,\alpha)$ is given in Eq. (2.31) by:

$$\left[\mu_r'\right]_{HB} = E(Y^r) = \frac{m^r ef_{\nu+r/2}(\alpha)}{ef_\nu(\alpha)} \,.$$

In order to obtain relationships between the moments of HB distribution, we use the following recurrence property of the function $ef_\nu(.)$ for $\nu > 0$ (Eq. (B.7), Appendix B):

$$ef_{\nu+1}(\alpha) = \frac{\alpha}{2} ef_{\nu+1/2}(\alpha) + \nu ef_\nu(\alpha) \tag{4.21}$$

The proposed approach is mainly based on Eq. (2.31). A general form of Eq. (4.21) can be deduced by recurrence and is given by:

$$ef_{\nu+r}(\alpha) = \frac{\alpha}{2} ef_{\nu+r-1/2}(\alpha) + (\nu+r-1) ef_{\nu+r-1}(\alpha) \tag{4.22}$$

Equation (4.22), combined to the non-central moments with Eq. (2.31) of HB, obtains:

$$E\left(\frac{X^{2r}}{m^{2r}}\right) = \frac{\alpha}{2} E\left(\frac{X^{2r-1}}{m^{2r-1}}\right) + (\nu+r-1) E\left(\frac{X^{2r-2}}{m^{2r-2}}\right), \quad (\nu+r) > 1 \tag{4.23}$$

An iterativeuse of this equation with $r = (1/2); 1; (3/2)$ leads to

$$E\left(\frac{X}{m}\right) = \frac{\alpha}{2} + \left(\nu - \frac{1}{2}\right) E\left(\frac{X^{-1}}{m^{-1}}\right) \tag{4.24a}$$

$$E\left(\frac{X^2}{m^2}\right) = \frac{\alpha}{2} E\left(\frac{X}{m}\right) + \nu \tag{4.24b}$$

$$E\left(\frac{X^3}{m^3}\right) = \frac{\alpha}{2} E\left(\frac{X^2}{m^2}\right) + \left(\nu + \frac{1}{2}\right) E\left(\frac{X}{m}\right) \tag{4.24c}$$

with $\nu > 1/2$.

The proposed approach corresponds to the generalized method of moments (Bobée and Ashkar, 1988). In fact, several moments equations could be combined to estimate the parameters.

By using Eqs. (4.24a) and (4.24b), we can obtain:

$$m^2 = \frac{2\text{Var}(X)}{2\nu\left[1 - E(X)E(X^{-1})\right] + E(X)E(X^{-1})} \tag{4.25}$$

Combining this equation with Eqs. (4.24b) and (4.24c) leads to:

$$\left(E(X^2)\right)^2 - E(X^3)E(X) = m^2\left[\nu E(X^2) - \left(\nu + \frac{1}{2}\right)\left(E(X)\right)^2\right] \tag{4.26}$$

By substituting Eq. (4.25) into Eq. (4.26), we can get:

$$\nu = \frac{E(X)E(X^{-1})\left[E(X^3)E(X) - \left(E(X^2)\right)^2\right] - \text{Var}(X)\left(E(X)\right)^2}{2\left[\left[1 - E(X)E(X^{-1})\right]\left[\left(E(X^2)\right)^2 - E(X^3)E(X)\right] - \left(\text{Var}(X)\right)^2\right]} \tag{4.27}$$

Note that this equation involves four non-central moments of order -1, 1, 2, and 3.

Concerning the estimation of the parameter α, we multiply Eqs. (4.24a) and (4.24b) by convenient terms and then we take their difference to obtain:

$$\alpha = \frac{m\left[2v\left(E(X)-E(X^2)E(X^{-1})\right)+E(X^2)E(X^{-1})\right]}{\mathrm{Var}(X)} \tag{4.28}$$

In Eqs. (4.25), (4.27), and (4.28), the theoretical moments are generally unknown. In order to obtain the estimators, those moments should be replaced by their estimates from the sample $x_1,...,x_n$ of size n, i.e., $E(X^2)$ is replaced by $n^{-1}\sum_{i=1}^{n} x_i^r$.

Thus, the MM estimation procedure of the three parameters of the HB distribution consists of three steps. First an estimator $\hat{v}$ of v is computed from Eq. (4.27); then by using Eq. (4.25) an estimator $\hat{m}$ of m is given; and finally, the estimation $\hat{\alpha}$ of α is obtained by substituting m and v with their respective estimators $\hat{m}$ and $\hat{v}$ in Eq. (4.28).

4.3.3. MM for HIB Distribution

In this subsection, we present the method of moments for Halphen type Inverse B distribution (HIB).

To obtain the MM parameter estimators of HIB distribution, we can follow the same steps as for HB distribution, by using Eq. (4.22) combined with HIB non-central moment Eq. (2.47). Alternatively, the relationship between a variable X from HB, and Y from HIB, can be used to obtain the estimators of HIB, directly from those of HB (cf Section 2.4.1, Eq 2.44). The estimator expressions m, α, and v for HIB, can be directly obtained by the transformation of X^r into Y^{-r} and m into m^{-1}, in the formulae in Eqs. (4.25), (4.27), and (4.28), given for HB. Formally, the explicit formulae of the three-parameter estimators are:

$$v = \frac{E(Y)E(Y^{-1})\left[E(Y^{-3})E(Y^{-1})-\left(E(Y^{-2})\right)^2\right]-Var(Y^{-1})\left(E(Y^{-1})\right)^2}{2\left[\left[1-E(Y)E(Y^{-1})\right]\left[\left(E(Y^{-2})\right)^2-E(Y^{-3})E(Y^{-1})\right]-\left(Var(Y^{-1})\right)^2\right]} \tag{4.29}$$

$$m^2 = \frac{2v\left[1-E(Y)E(Y^{-1})\right]+E(Y)E(Y^{-1})}{2\mathrm{Var}(Y^{-1})} \tag{4.30}$$

$$\alpha = \frac{2v\left[E(Y^{-1})-E(Y^{-2})E(Y)\right]+E(Y^{-2})E(Y)}{m\mathrm{Var}(Y^{-1})} \tag{4.31}$$

In Eqs. (4.29), (4.30), and (4.31), the theoretical moments are generally unknown. In order to obtain the estimators, those moments should be replaced by their estimates from the sample $x_1,...,x_n$ of size *n*, i.e., $E(X^r)$ is replaced by $n^{-1}\sum_{i=1}^{n} x_i^r$. As in the case of HB, the parameter ν is estimated firstly (Eq. 4.29) then *m* and α are obtained from Eqs (4.30) and (4.31), respectively.

4.4. The ITERATIVE MIXED METHOD (IMM)

In order to accelerate numerical resolution of the maximum likelihood system, the proposed method reduces the variation interval that contains the optimal value of ν that maximizes $\ln L(\nu \mid \alpha, m)$. Therefore, from a theoretical point of view, this method is as optimal as the ML method, but less expensive in terms of computational time (Chebana et al. 2010). It is another numerical way to resolve the same ML system.

In the ML method, the parameter ν is varied within a predefined interval (cf. Section 4.2), namely:

- For HA: $\nu \in \left]-U;U\right[$ with $U = AH^{-1}/\left[AH^{-1}-1\right]$;
- For HB: $\nu \in \left]0;V\right[$ with $V = 1/\left[2\left(Q/A^2-1\right)\right]$
- For HIB: $\nu \in \left]0;W\right[$ with $W = 1/\left[2\left(H^2IQ^{-1}-1\right)\right]$.

In the IMM, the estimator ν_{MM} (obtained in Section 4.3) is used as an initial value to the iterative process. Then, the variability direction of ν is determined according to the sign of the derivative of the partial log-likelihood function. From the initial point, the values of ν are increased by a fixed step (positive or negative), and we evaluate the partial log-likelihood function after having estimated α and *m*. The change of sign of the derivative of the log-likelihood function, or the sign of the difference $\Delta_k - \Delta_{k-1}$ (see algorithm below), indicates stopping of iterations and hence, obtaining the maximum of the log-likelihood function. For example for HA, Figure 4.1 illustrates the variation range of ν, corresponding to this method and that of the ML method.

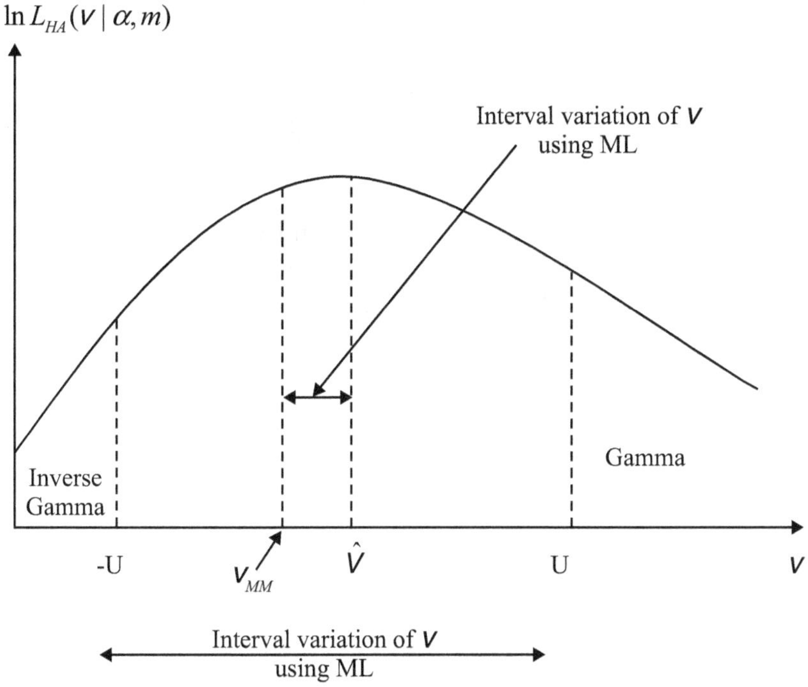

Figure 4.1: Illustration of interval variations of v using ML and IMM for HA distribution.

More explicitly, the following algorithm shows how this method works:

1. Select a step s: for example take the same value used in ML, that is $s = 0.1$. This value is generally convenient for situations encountered in hydrology.
2. Initial stage:
   a. Obtain $\hat{v}_0 = v_{MM}$: using Eqs. (4.20), (4.27), and (4.29) for HA, HB, and HIB, respectively.
   b. Compute $\hat{\alpha}_0$ and $\hat{m}_0$: by substituting the corresponding $\hat{v}_0$ in Eqs. (4.5a,b) for HA; Eqs. (4.9a,b) for HB; and Eqs. (4.13a,b) for HIB.
   c. Compute the partial log-likelihood function $L_0 = \log L(\hat{v}_0 \mid \hat{\alpha}_0, \hat{m}_0)$ using Eqs. (4.7), (4.11), or (4.15), according to the type of the Halphen distribution.

3. Stage 1: take $\hat{v}_1 = \hat{v}_0 + s$ and

    a. Compute $\hat{\alpha}_1$ and $\hat{m}_1$ that correspond to $\hat{v}_1$ using the same equations as in stage Eq. (2.b).

    b. Compute the partial log-likelihood function $L_1 = \log L(\hat{v}_1 \mid \hat{\alpha}_1, \hat{m}_1)$, as in Eq. (2.c).

    c. Compute the difference between L_1 and L_0, denoted by $\Delta_1 = L_1 - L_0$.

4. Stage k $(k = 2,3,\ldots)$:

    a. If Δ_1 is positive:

       While $\Delta_k = L_k - L_{k-1}$ is positive, take $\hat{v}_k = \hat{v}_{k-1} + s$ and the corresponding $\hat{\alpha}_k$ and $\hat{m}_k$ to evaluate $L_k = \log L(\hat{v}_k \mid \hat{\alpha}_k, \hat{m}_k)$.

    b. If Δ_1 is negative:

       While $\Delta_k = L_k - L_{k-1}$ is negative, then take $\hat{v}_k = \hat{v}_{k-1} - s$ and the corresponding $\hat{\alpha}_k$ and $\hat{m}_k$ to assess $L_k = \ln L(\hat{v}_k \mid \hat{\alpha}_k, \hat{m}_k)$.

5. Stop iterations when the sign of Δ_k changes. That means the maximum is reached.

6. Finally, the obtained estimators are $\hat{v}_{MMI} = \hat{v}_{k^*-1}$, $\hat{\alpha}_{MMI} = \hat{\alpha}_{k^*-1}$ and $\hat{m}_{MMI} = \hat{m}_{k^*-1}$, where k^* is associated to the final step, i.e. it corresponds to the sign change of the derivative of the partial log-likelihood function.

Since the variation range of v is reduced in this method, it is possible to come as close as we want to to the optimum by taking the smaller value of step s. Note that for comparison purposes, the value of s is taken to be the same for both ML and IMM.

When implementing the algorithm of the IMM method for each Halphen distributions, there are three elements that need to be pointed out:

- Although the estimator ν_{MM} belongs to the variation interval of ν, it is not necessarily one of the subdivision values employed in ML. It is likely that ν_{MM} can be found between two points of this grid.
- In the IMM algorithm, the value of ν of the last iteration is not at the maximum of the partial log-likelihood function. It corresponds to the change of sign variations of that function. Hence, the maximum of the partial log-likelihood function corresponds to the solution given by the penultimate iteration.
- The iteration number is finite (and small), since the partial log-likelihood of Halphen distributions are strictly concave (Cf. Section 4.2).

4.5. The BAYESIAN APPROACH FOR HALPHEN DISTRIBUTIONS

In the last section, we considered the maximum likelihood and the moment methods for determining estimators for the vector of parameters $\underline{\theta}=(m,\alpha,\nu)$ of the Halphen distributions. The bayesian framework treats parameters as random variables that can be described by probabilistic statements (Robert, 2006). A brief introduction to the bayesian framework is given in Appendix D. The first recommendations to the use of the bayesian approach in hydrology are given in the pioneering work of Bernier (1967). A methodological procedure to combine different sources of information in a Bayesian framework is presented in Bernier (1998) to tackle some classical problem in statistical hydrology.

The use of Halphen distribution in bayesian framework (Appendix D) is very limited in the literature. El Adlouni (2002) proposed a Markov Chain Monte Carlo (MCMC) algorithm for subjective priors $\pi(\cdot)$, which takes into account the supports of the parameters of HA distribution. In the case of the HA and HB distributions, Parent et al. (2012) suggested the use of independent improper prior for inverse of the scale parameter and independent uniform priors with wide supports for the shape parameters. Here is a summary of their approach, where the following parameterization had been considered:

$$\rho = 1/m$$

Then the log-likelihood functions of the $HA(m,\alpha,v)$ and $HB(m,\alpha,n)$ distributions are given respectively by:

$$LogL_{HA} = -\alpha\left(\rho T_2^{HA} + \frac{T_3^{HA}}{\rho}\right) + (v-1)T_1^{HA} - n\log\left(2K_v(2\alpha)\right) + vn\log\rho \tag{4.32}$$

$$LogL_{HB} = -T_3^{HB}\rho^2 + \alpha T_2^{HB}\rho + (2v-1)T_1^{HB} - n\log ef_v(\alpha) + n\log 2 + 2vn\log\rho \tag{4.33}$$

where

- $T_1^{HA} = \ln G$
- $T_2^{HA} = A$
- $T_3^{HA} = H^{-1}$

are the sufficient statistics of the HA distribution and are given by Eq. (2.24); and,

- $T_1^{HB} = \ln G$
- $T_2^{HB} = Q$
- $T_3^{HB} = A$

are the sufficient statistics of the HB distribution and are given by Eq. (2.42).

Parent et al. (2012) suggested the use of independent non-informative prior for the scale parameter m and uniform independent distributions with large supports for the shape parameters α and v. For an observed sample $\underline{x}$ from HA and $\underline{y}$ from HB, the posterior distributions of the parameters of the HA and HB distributions are given, respectively, by:

$$\pi\left(\rho,\alpha,v\middle|\underline{x}\right) \propto \left(\frac{1}{K_v(2\alpha)}\right)^n \exp\left(-\alpha\left(\rho T_2^{HA} + \frac{T_3^{HA}}{\rho}\right) + (v-1)T_1^{HA} + (vn-1)\log\rho\right) \tag{4.34}$$

$$\pi\left(\rho,\alpha,v\middle|\underline{x}\right)\propto\left(\frac{1}{K_v\left(2\alpha\right)}\right)^n\exp\left(\begin{matrix}-\alpha\left(\begin{matrix}\rho T_2^{HA}+\\ \dfrac{T_3^{HA}}{\rho}\end{matrix}\right)+\\ \left(v-1\right)T_1^{HA}+\left(vn-1\right)\log\rho\end{matrix}\right) \tag{4.35}$$

Then the conditional posterior distributions for the scale parameter ρ are HA or HB with the parameters (Parent et al. 2012):

$$\pi\left(\rho\middle|\underline{x},\alpha,v\right)\equiv HA\left(\frac{T_3^{HA}}{T_2^{HA}},\alpha\sqrt{T_2^{HA}T_3^{HA}},vn\right) \tag{4.36}$$

$$\pi\left(\rho\middle|\underline{y},\alpha,v\right)\equiv HB\left(\frac{1}{\sqrt{T_3^{HB}}},\alpha\frac{T_2^{HB}}{\sqrt{T_3^{HB}}},vn\right) \tag{4.37}$$

Then the explicit form of the joint posterior marginal of α and v can be deduced from Eqs. (4.36) and (4.37) and are given by (Parent et al. 2012):

For HA

$$\pi\left(\alpha,v\middle|T^{HA}\right)\propto\frac{m_n^{v_n}K_{v_n}\left(2\alpha_n\right)}{K_v^n\left(2\alpha\right)}\exp\left[\left(v-1\right)T_1^{HA}\right] \tag{4.38}$$

For HB

$$\pi\left(\alpha,v\middle|T^{HB}\right)\propto\frac{m_n^{v_n}ef_{v_n}\left(\alpha_n\right)}{ef_v^n\left(\alpha\right)}\exp\left[\left(2v-1\right)T_1^{HB}\right] \tag{4.39}$$

The bayesian inference is based on the conditional posterior distribution. For given loss function, quadratic or 0–1, the bayesian estimator will corresponds to the mean or the mode of the posterior distribution. Given that no explicit form is available for the all posterior distribution of the parameters of the Halphen distributions, simulation methods such as Markov Chain Monte Carlo (MCMC) (cf. Section 5.4) could be considered to approximate its mode or median. However, more analysis is needed to select the most appropriate instrumental distributions.

4.6. CONCLUSION

This present chapter covered parameter estimation of the three Halphen distributions with classical and new mixed estimation methods.

Theoretically, the ML method is optimal. Nevertheless, to resolve the corresponding non-linear system, numerical approximations are necessary. Moreover, the maximization of the partial log-likelihood function requires a significant number of iterations. Furthermore, the bias of the ML is asymptotically null but it may be large for small sample sizes. Hence, it is relevant to consider other estimation methods for small sample sizes frequently encountered in hydrology. However, the method of moments (MM, Section 4.3) gives estimators that have explicit forms. Despite that it leads to good results, the MM approach is not optimal in theory, since it is not based on the sufficient statistics of Halphen distributions. These reasons motivated the development of the iterative mixed methods IMM to estimate a near optimum solution, combining the advantages of MM (explicit formulae) and ML (optimality of the estimators) methods to get the optimal values.

A performance comparison of the three estimation methods (MM, ML, and IMM) for the three Halphen distributions was conducted using Monte Carlo simulations (Chebana et al. 2010). Thus comparison shows homogeneity of the behavior of all compared methods. The comparison, in terms of quantiles, shows a great similarity of the results of the three methods for the HA distribution; however, the IMM method is the most adequate in both theoretical and empirical point of view (Section 4.4). Nevertheless, in practice and with the same performance, MM method may be recommended as well. For HB distribution, the MM method can be generally recommended to estimate quantiles of large return periods. However, in the case of large skewness in the range of HB ($C_s \approx 0.8$; Figure 2.5) the IMM method may be preferable. For HIB distribution, the IMM method can be recommended considering results and computation time, except in the case of a very low skewness in the range of HIB ($C_s \approx 0.8$; Figure 2.6) where the MM method is recommended. Finally, it can be stated that the IMM method can be considered for a majority of cases encountered in hydrology. The main idea of the IMM could be applied to other distributions with several parameters and complicated forms, especially for distributions of the exponential family.

Although, it has been proposed along time ago by pioneering work of Bernier (1967), the bayesian approach is very promising. The previous researches, cited in this chapter, are good avenues to pursue. Thanks to computational tools such as MCMC algorithms (cf. Section 5.4), this approach may be used to combine several sources of information. Recent development of the MCMC algorithms could lead to unbiased estimators when the instrumental distributions are well chosen.

Chapter 5
Monte Carlo Simulations for Halphen Distributions

5.1. INTRODUCTION

Despite the availability of the numerical techniques for parameter estimation, many other computational problems, related to statistical inference with Halphen distributions, need a sophisticated approach. For example:

a) Evaluate estimator properties such as: bias, sample variance for the quantiles, empirical distribution, and deduce a bootstrap confidence interval. In fact, although Halphen distribution parameter estimators are efficient, no study was done, before 2007 (El Adlouni and Bobée, 2007), to assess bias estimators in the case of finite sample sizes.

b) Check if a procedure for constructing a confidence interval for parameters and quantiles achieves the specified nominal level of coverage. Confidence intervals, given by the software HFG for Halphen distributions, are asymptotic. It is, thus, important to study their properties in practice for finite sample sizes.

c) Verify if a hypothesis testing procedure reaches the fixed level. In fact, statistics of some goodness-of-fit tests are not pivotal (depending on null hypothesis). Thus, critical regions, corresponding to fixed level of significance, depend on the null hypothesis and can lead to complicated formulas for the Halphen distributions.

All these computational problems, presented above can be resolved using the Monte Carlo (MC) techniques. MC methods encompass any technique of statistical sampling employed to approximate solutions to quantitative problems. The MC methods provide approximate solutions to a variety of mathematical problems by performing statistical sampling experiments on a computer. For example, let X be a random variable with distribution function F and probability density function f. The T-year event, x_T, corresponds to (1–1/T)-quantile of X and is such that: $\int_{-\infty}^{x_T} f(x)\,dx = 1-(1/T) = F(x_T)$ which is equivalent to the Eq. (1.2)

$x_T = F^{-1}(1-(1/T))$; where $F^{-1}(p) = \inf\{x : F(x) \geq p\}$. When F is not invertible in explicit form, MC methods are generally used to approximate the T-year event. Their efficiency depends on the availability of techniques to draw samples from the distribution function. The fact that Halphen distribution functions are not invertible, in a closed form, makes the use of direct method (based on uniform random number generators) impossible. However, other MC techniques can be used to resolve these computational problems related to the use of Halphen distributions in statistical inference, such as the Markov Chain Monte Carlo (MCMC) and the Importance Sampling (IS) methods (Robert, 2006). For more detail on these last methods and a comparative study see (El Adlouni and Bobée, 2007).

5.2. MONTE CARLO SIMULATION AND RANDOM VARIABLE GENERATION

Monte Carlo methods encompass any technique of statistical sampling employed to approximate solutions to quantitative problems. The MC methods are indicated in cases where the dimensionality and/or complexity of a problem make straightforward numerical solutions impossible or impractical. This technique performs particularly well when the process is one where the underlying probabilities are known but lead to intricate systems. The most common uses of Monte Carlo techniques are related to the assessment of estimator properties, tests theory and approximation of integrals (Rubinstein, 1981). Therefore it constitutes a practical alternative for quantile approximation in the case of a complex pdf.

The applicability of Monte Carlo techniques depends on the availability of an efficient procedure to generate samples from the distribution of interest. The well-known general methods to generate samples from non-uniform distributions are (Devroye, 1986):

- The inversion method: when the inverse of the cumulative distribution function $X(F)$ can be expressed explicitly by Eq. (1.2),
- The composition method: when it is possible to write the cdf of interest as a combination of elementary distributions,

- The Acceptance-Rejection method (A-R): when the cdf is not invertible in a closed form, but its statistical properties are well known.

This last method, used in this work, is based on the existence of an instrumental distribution with specific characteristics. The knowledge of the mathematical and statistical properties of Halphen distributions makes the application of the A-R method straightforward. The following section presents techniques to choose the instrumental distributions for efficient A-R procedures to draw samples from Halphen distributions.

5.3. THE ACCEPTATION-REJECTION APPROACH FOR THE HALPHEN DISTRIBUTIONS

5.3.1. The Acceptance-Rejection Method (A-R)

The Acceptance-Rejection (A-R) method is a sample generating method mainly used for continuous random variables (Devroye 1986). Suppose *f*, is the probability density function (pdf) from which to generate the sample distribution; and *g* is the pdf of a well-known distribution with an available algorithm of generation. Moreover, suppose that we can find a scaling factor *M* which holds the following inequality:

$$f(x) \leq M\,g(x) = e(x) \tag{5.1}$$

Thus *e* is an envelope whose graph lies completely above the graph of f. A condition that is necessary to obtain Eq. (5.2) is:

$$Supp(f) \subset Supp(g) \tag{5.2}$$

Where $Supp(f)$ corresponds to the support of the distribution f.

Under this condition, the Acceptance-Rejection algorithm, to generate samples from distribution f, is given by the following steps:

1. Generate x from g distribution.
2. Generate u from the uniform distribution $U_{[0,1]}$.
3. Accept x if $u \leq \dfrac{f(x)}{Mg(x)}$, otherwise go to step 1.

The efficiency of an A-R algorithm depends on the scaling factor M in Eq. (5.1) which dictates the average number of iterations needed for a proposed candidate x to be accepted. In order to have an effective algorithm, M must be as small as possible. Thus it is necessary to find an adequate instrumental distribution to check the condition of Eq. (5.2) and to verify the smallest factor M in Eq. (5.1). The following subsections, present the implementation of the A-R method to generate samples from Halphen distributions (El Adlouni and Bobée, 2007).

5.3.2. Halphen Type A Distribution (HA)

We present here the A-R algorithm to generate samples from a random variable X with pdf f_{HA} given by Eq. (2.2). The Gamma distribution is chosen as instrumental distribution, because it is a limiting case of the HA distribution and has a simple and efficient generation algorithm.

For a random variable Y with Gamma distribution (Gam), the pdf g is Eq. (2.25):

$$g(y;\alpha,\lambda)=\frac{\alpha^{\lambda}}{\Gamma(\lambda)}y^{\lambda-1}\exp\left[-\alpha y\right] \quad , \quad y>0$$

where $\alpha(>0)$ is the scale parameter and $\lambda(>0)$ is the shape parameter.

The mean and the variance of Y are given by, respectively:

$$\begin{cases} E\left[Y\right]=\dfrac{\lambda}{\alpha} \\ Var\left[Y\right]=\dfrac{\lambda}{\alpha^2} \end{cases} \tag{5.3}$$

The parameters of the Gamma distribution should be chosen such as the condition in Eq. (5.2), which is verified by $(\text{i.e.}Supp(f_A)\subset Supp(g))$. To ensure this condition, we consider the Gamma distributed random variable Y (Ahrens and Dieter, 1974):

$$\begin{cases} E\left[Y\right]=E\left[X\right] \\ Var\left[Y\right]=2Var\left[X\right] \end{cases} \tag{5.4}$$

where $E[X]$ and $Var[X]$, mean and variance of HA distribution, are given by Eqs. (2.11) and (2.15a), respectively. Thus, the parameters of the Gamma distribution can be given as a function of HA moments using Eqs. (5.3) and (5.4); then as a function of the HA parameters using Eqs. (2.11) and (2.15a):

$$\lambda = \frac{\{E[X]\}^2}{2Var[X]} \quad \text{and} \quad \alpha = \frac{\lambda}{E[X]} \tag{5.5}$$

Figures 5.1 and 5.2 present pdf of HA, Gamma, and the envelope e (Eq. (5.1)). They show that with this choice of the Gamma parameters, the support of g contains that of f_A (condition given in Eq. (5.2)).

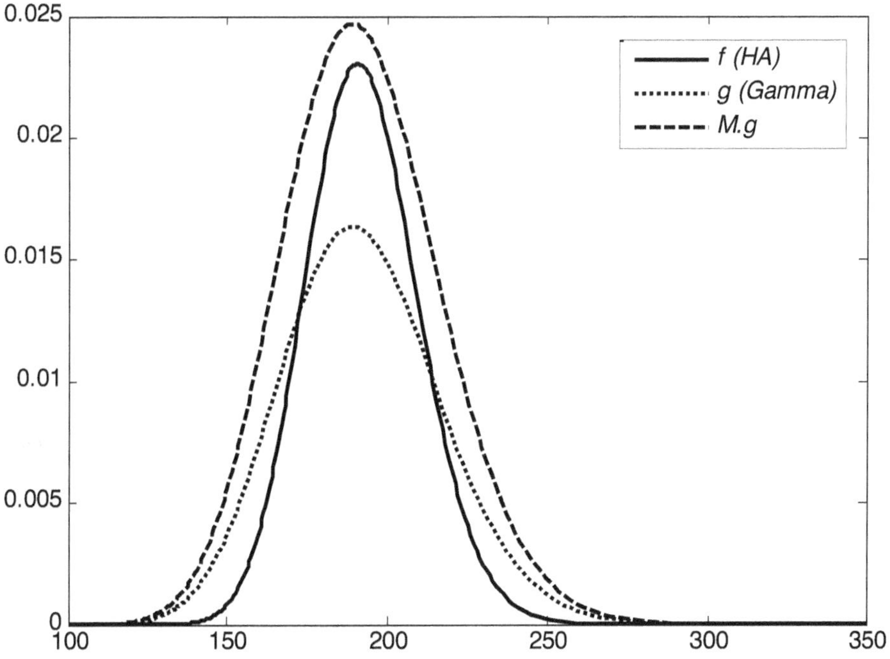

Figure 5.1: pdf of HA (Case 1, Table 5.1) and instrumental distribution (Gamma) with the envelope *e* of A-R algorithm.

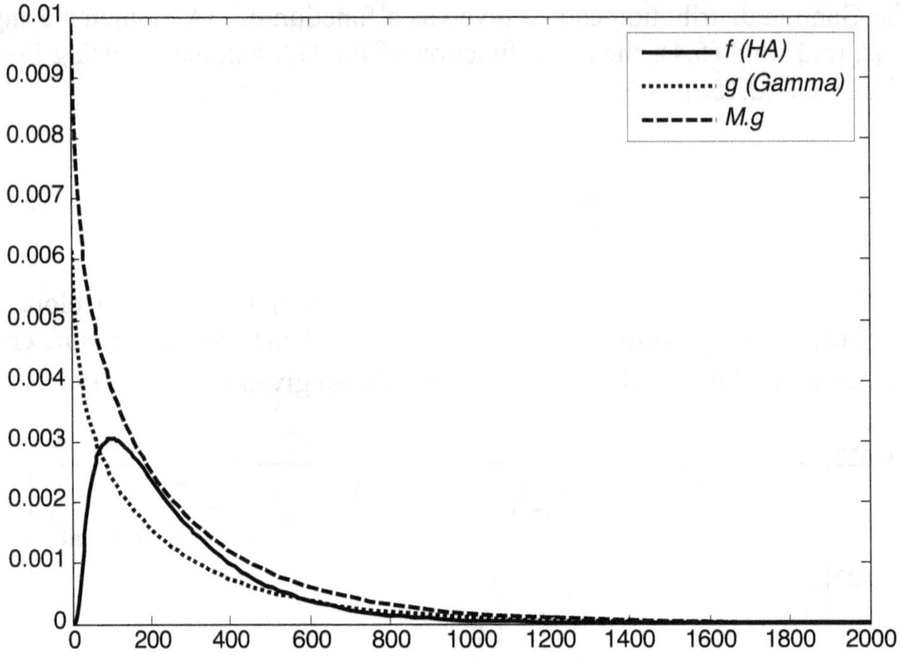

Figure 5.2: pdf of HA (Case 5, Table 5.1) and instrumental distribution (Gamma) with the envelope *e* of AR algorithm.

These two cases are presented among five examples in Table 5.1, as studied in El Adlouni and Bobée (2007). The scale factor M in Eq. (5.1) can be obtained by a numerical maximization of $(f_A(x))/(g(x))$ with the support of f_A .

Table 5.1: Characteristics of the HA distributions and corresponding Gamma distributions.

	HA					Gamma		A-R
	ν	**A**	**m**	C_v	C_s	α	$1/\lambda$	**M**
Case 1	**70.00**	**50.00**	**100**	0.09	0.22	**61.15**	**3.14**	1.51
Case 2	**10.90**	**3.20**	**100**	0.27	0.59	**6.38**	**57.96**	1.52
Case 3	**2.50**	**2.95**	**100**	0.38	1.00	**3.31**	**47.63**	1.55
Case 4	**5.00**	**8.00**	**100**	0.66	1.40	**1.12**	**474.24**	1.56
Case 5	**1.00**	**0.50**	**100**	0.78	1.80	**0.80**	**334.19**	1.61

The five cases are chosen from observed hydrological dataset to cover the region corresponding to the HA distribution in the (C_v, C_s) diagram (Figure 2.4).

5.3.3. Halphen Type B Distribution (HB)

The pdf of the HB distribution has various forms according to the values of the parameters. It can be without mode, it can have a unique mode or have one mode and an anti-mode (Perreault et al. 1999a, cf. 2.3.1). These three cases (Figure 2.2) occur under the following conditions.

1. The pdf of the HB distribution allows no mode if $\alpha < 0$ and $\nu < 0.5$ or $(-\alpha/4)^2 \le 0.5 - \nu$; are verified;
2. It has one mode if $\alpha \ne 0$ and $\nu \ge 0.5$;
3. It possesses one mode and an anti-mode if $\alpha \le 0$, $\nu < 0.5$, and $(-\alpha/4)^2 > 0.5 - \nu$.

The procedure presented for the HA distribution is used again to generate samples from the HB distribution, using the A-R method. Simulations show that the various form of HB distribution can be summarized by two cases: $\nu \le 1$ or $\nu > 1$ (El Adlouni and Bobée, 2007). The first one corresponds to HB distributions with no mode or with a mode and an anti-mode. The second case is observed when the pdf function has only one mode. When $\nu \le 1$, the pdf of the HB distribution, has a heavier tail than that of the case $\nu > 1$. Therefore it is necessary to consider an instrumental distribution with a large support to verify the condition (Eq. (5.2)). For $X \sim HB(m,\alpha,\nu)$, the following instrumental distribution is considered: $Y \sim Gam(\alpha,\lambda)$.

Case 1 $\nu \le 1$

The parameters α and λ are such as:

$$\begin{cases} E\left[Y\right] = E\left[X\right] \\ Var\left[Y\right] = 5Var\left[X\right] \end{cases} \tag{5.6}$$

Case 2 $\nu > 1$

The parameters α and λ are such as:

$$\begin{cases} E[Y] = E[X] \\ Var[Y] = 2Var[X] \end{cases} \tag{5.7}$$

The multiplicative factor in the variance expression are chosen by intensive simulation studies. The adequacy of the choice of instrumental distribution is illustrated in Figures 5.3 and 5.4, for two cases among the five cases show in Table 5.2. This distribution was taken into account in Section 5.4 for comparison purposes. These figures show that (1) the support generated from the HB distribution is contained in that of the instrumental distribution (condition in Eq. (5.2)); and (2) the envelope $e = M.g$ verifies the inequality Eq. (5.1). With this choice, the support of the HB distribution is contained in the Gamma distribution (Figures 5.3 and 5.4), even for a high value of the coefficient of skewness C_s.

As in the case of the HA distribution, parameters of the instrumental distributions for HB are calculated using Eqs. (5.6) and (5.7).

Table 5.2: Characteristics of the HB distributions and corresponding Gamma distributions.

	HB					Gamma		A-R
	ν	**A**	**m**	C_v	C_s	α	$1/\lambda$	**M**
Case 1	**0.60**	**2.90**	**100**	0.42	0.20	**1.11**	**139.23**	2.51
Case 2	**0.50**	**1.40**	**100**	0.61	0.59	**0.52**	**172.93**	4.64
Case 3	**0.30**	**1.00**	**100**	0.85	0.99	**0.27**	**214.11**	4.62
Case 4	**0.20**	**0.70**	**100**	1.10	1.44	**0.16**	**238.82**	4.16
Case 5	**0.17**	**0.30**	**100**	1.28	1.80	**0.12**	**233.30**	5.48

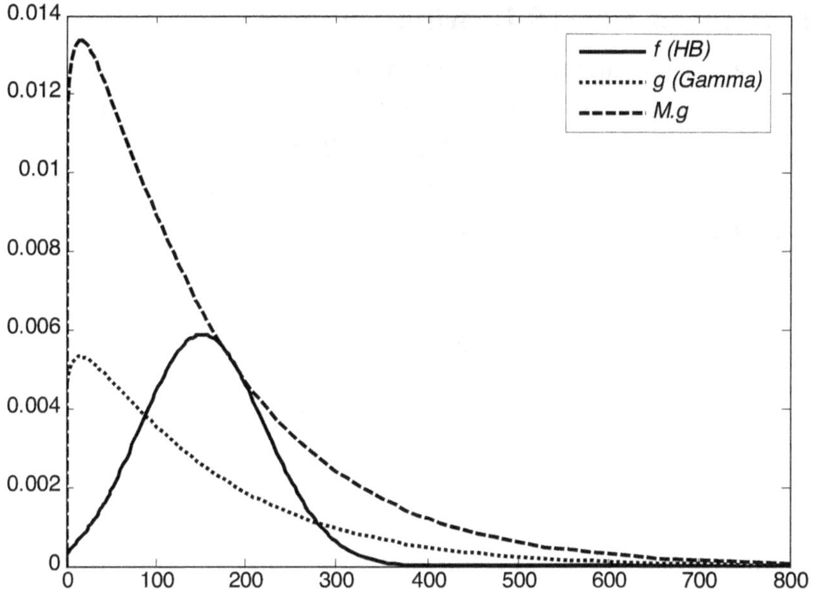

Figure 5.3: pdf of HB (Case 1, Table 5.2) and instrumental distribution (Gamma) with the envelope *e* of A-R algorithm.

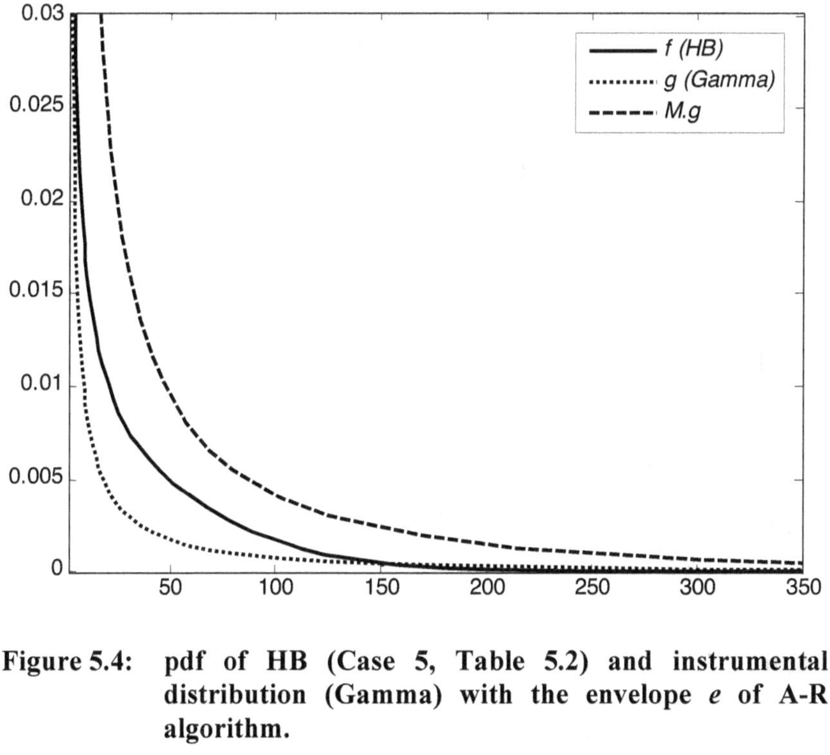

Figure 5.4: pdf of HB (Case 5, Table 5.2) and instrumental distribution (Gamma) with the envelope *e* of A-R algorithm.

5.3.4. Halphen Type IB Distribution (HIB)

The A-R algorithm for the HB distribution, presented in Figures 5.3 and 5.4, can be used to deduce samples generated from the HIB. If $X \sim HB\left(x;m,\alpha,\nu\right)$ then: $Y = 1/X \sim HIB(y;m^{-1},\alpha,\nu)$ (cf. Section 2.4, Eq. (2.44)). Using this relationship, we can draw samples from $HIB(\mathrm{y};m,\alpha,\nu)$ distributions, by inversion of those generated from $HB(1/y;m^{-1},\alpha,\nu)$.

For a comparison study, we present the direct A-R algorithm to generate samples from HIB distribution, with the same instrumental distribution as in the case of HA and HB distributions. Note that, for the A-R algorithm, the most important condition is given by the inequality in Eq. (5.1). So any distribution, which verifies this condition, can be considered in the generation process. As already mentioned, the Gamma distribution was considered for two reasons: the availability of effective algorithms for its generation and the availability of simple methods to estimate parameters according to the first and second moment.

The Gamma distribution considered in the case of the HIB has the same mean and double variance. Simulation studies show that with this choice, the condition in Eq. (5.2) is verified as show in Figures 5.5 and 5.6.

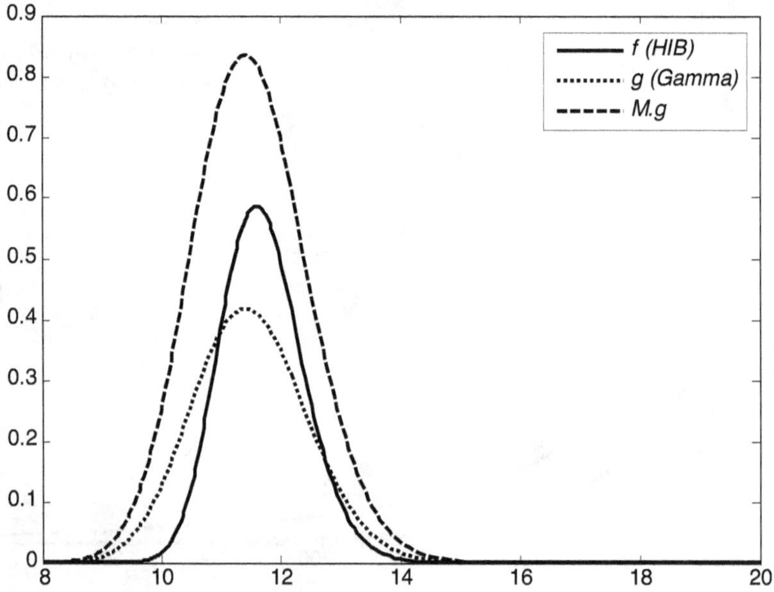

Figure 5.5: pdf of HIB (Case 1, Table 5.3) and instrumental distribution (Gamma) with the envelope *e* of A-R algorithm.

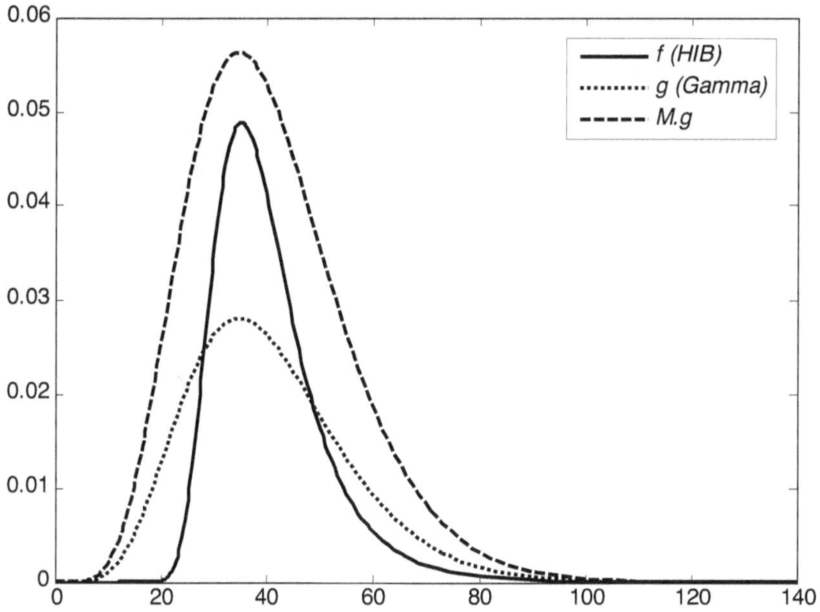

Figure 5.6: pdf of HIB (Case 5, Table 5.3) and instrumental distribution (Gamma) with the envelope *e* of A-R algorithm.

Table 5.3 presents the HIB distributions used to check A-R conditions and for comparison studies. It is pointed out that this direct approach is presented for comparative purpose (cf. Section 5.4) and can be applied only if $\nu > 1$, i.e. when the HIB distribution has a finite variance. If $\nu \leq 1$, the variance is infinite and the direct method cannot be used. However, the first approach, i.e. by inverting sample generated from corresponding HB distribution, can be used with any value of ν.

Table 5.3: Characteristics of the HIB distributions and corresponding Gamma distributions.

	HIB					Gamma		A-R
	ν	A	m	C_v	C_s	α	$1/\lambda$	M
Case 1	**72.00**	**0.44**	**100**	0.06	0.29	**143.61**	**0.08**	2.00
Case 2	**11.30**	**8.60**	**100**	0.10	0.60	**45.63**	**0.36**	1.71
Case 3	**1.30**	**9.50**	**100**	0.15	1.00	**21.93**	**0.94**	1.81
Case 4	**3.10**	**5.00**	**100**	0.21	1.39	**11.14**	**2.81**	1.90
Case 5	**3.00**	**3.20**	**100**	0.26	1.79	**7.16**	**5.66**	2.01

5.4. MARKOV CHAIN MONTE CARLO METHOD (MCMC)

Markov Chain Monte Carlo (MCMC) is a simulation technique that allows samples to be drawn from complex and high dimensional probability distributions. The main objective of the MCMC method (Hastings, 1970) is to generate values of a random vector X with probability distribution f, and to calculate Monte Carlo approximations of any quantity expressed by an integral, generally with no analytical solution. In order to do this, an ergodic Markov chain with stationary distribution f is generated. Then, under mild conditions and after a large number of transitions, called the burn-in period of the chain, we obtain an approximate sample from f. Many MCMC algorithms are available in the literature, namely the Metropolis-Hastings algorithm and the Gibbs sampler (Metropolis et al. 1953; Hastings 1970; Geman and Geman 1984; and Gelfand and Smith 1990). Due to its easy implementation, many statistical applications have considered MCMC methods to solve computational and optimization problems. MCMC methods are usually used for complex distributions, when any statistical and mathematical properties are known. Here, MCMC methods were used for comparative purpose. For each type of the Halphen distributions we use the Metropolis-Hastings algorithm (MH) and compare its performance to the A-R method. Markov chain iterations are updated using a random-walk Metropolis algorithm with a Lognormal proposal density, centered at the current state of the chain. A brief description of the MH algorithm is presented below.

Metropolis-Hastings algorithm (MH): For a conditional density, $q(y|x)$, such that for each fixed x, it is straightforward to draw random variable from $q(.|x)$. The basic Metropolis-Hastings algorithm (MH) is given by the following steps:

1. Choose an arbitrary starting value $x^{(0)}$ and set $t=0$;
2. Sample an observation y from $q(.|x^{(t)})$;
3. Set $x^{(t+1)}=y$ with probability

$$\alpha_p\left(x^{(t)},y\right)=\min\left(1,\frac{f(y)}{f\left(x^{(t)}\right)}\frac{q\left(x^{(t)}\mid y\right)}{q\left(y\mid x^{(t)}\right)}\right)$$

otherwise reject y and set $x^{(t+1)} = x^{(t+1)}$, and put $t = t+1$;

4. if $t < N$, return to step 2, otherwise stop.

These steps are required for simulating a Markov chain $\{x^{(0)}, x^{(1)}, \ldots, x^{(N)}\}$ of length N. This algorithm is very flexible and can be adapted for almost any instrumental distribution q that has at least the same support as f. The normal distribution is usually used with the Metropolis-Hastings algorithm, to simplify the expression of acceptance probability α_p. The supports of Halphen distributions are positives. Thus, the Lognormal distribution, centered on the current state of the chain and with the same variance as the target distribution, is proposed:

$$q\left(y \middle| x^{(t)}\right) \equiv Lognormal\left(\mu, \sigma\right) \tag{5.8}$$

The parameters of the Lognormal distribution are chosen such as: $E(y \mid x^{(t)}) = x^{(t)}$ and $Var(y \mid x^{(t)}) = 2Var(X)$. X is the random variable needed to be generated, and is distributed as HA, HB, or HIB. This choice of the conditional distribution allows the Markov Chain to cover all the support of the distribution to generate. Thus, the parameters of the Lognormal distribution are:

$$\sigma = \sqrt{\log\left(\frac{Var\left(X\right)}{\left(x^{(t)}\right)^2} + 1\right)}$$

and

$$\mu = \log\left(x^{(t)}\right) - \frac{\sigma^2}{2} \tag{5.9}$$

In practical uses of MCMC methods, the length of the chain and the burn-in time should be determined to answer the following questions:

a) Starting from which iteration has the generated chain reached stationary state?

b) Is the chain long enough to cover the whole support of the distribution of interest?

c) Do the generated observations allow the adequate estimation of the parameters with a given precision?

Some methods to assess the convergence of MCMC methods make it possible to determine the length of the chain and the burn-in time (El Adlouni et al. 2006). Simulation studies show that, for Halphen distributions and using the MH algorithm as described above, stationary state is reached for all considered cases with chain's length $N = 60000$ and a burn-in time of $N_0 = 20000$. For more detail on the convergence of the MCMC algorithms, an exhaustive review with comparative study can be found in El Adlouni et al. (2006).

For the HB and HIB distributions, Parent et al. (2016) suggested the use of a mixture of the Metropolis-Hastings approach, and the Importance Sampling (IS) technique (Devroye, 1986), to avoid computing the Exponential factorial functions. This approach is detailed in Andrieu and Roberts (2009), and is based on a Metropolis–Hastings algorithm that replaces the intractable target density, with an unbiased estimator based on an importance sampling algorithm. The approach is termed pseudo-marginal, where a positive unbiased estimate of the posterior distribution is utilized, rather than evaluating the posterior exactly.

5.5. CONCLUSION

Halphen distributions constitute a complete system to model hydrological series. With their limiting forms: Gamma and Inverse Gamma distribution, the Halphen family of distributions allows the ability to fit a large variety of data sets. Halphen distributions differ from other three-parameter distributions used in hydrology. For each of the three distributions, the parameters can be estimated from a triplet of sufficient statistics. Thus, the ML estimators of Halphen distribution parameters are efficient (minimum variance) even for finite sample sizes, although a bias may exist. The generating procedures, A-R and MH, presented in this chapter make it possible to study parameter estimator properties such as bias and empirical distribution. In fact, even if maximum likelihood estimators of Halphen distribution parameters are efficient, their properties particularly the bias, should be evaluated, in the case of finite sample sizes. Comparison of generating methods, A-R and MH, show that A-R algorithms give good results when comparing empirical distributions of the generated samples to the theoretical Halphen distributions. These generating techniques can be used for constructing confidence intervals and to determine critical regions for parametric and goodness of-fit tests. In addition to the availability of computational tools to estimate parameters and quantiles (HFG software, cf. Chapter 6), Monte Carlo procedures presented in this chapter, allow for solving many computational problems encountered when studying Halphen distributions. Thus, all these techniques will be helpful and much more practical when using the Halphen distributions.

Chapter 6
Matlab codes for Halphen Fitting and Generating (HFG)

6.1. INTRODUCTION

In this chapter, the Matlab codes for Halphen Fitting and Generating (HFG) are presented. The main objective is to offer computational tools to make in depth research on Halphen family or to use these distributions in applied studies. The availability of these codes allows user-friendly operations by including these functions in other research studies.

The following sections introduce Matlab codes for:

- The Decision Support System (DSS, Section 3.3.3);
- the generation of datasets, from each of the three Halphen distributions, for given values of the parameters (cf. Chapter 5);
- the method of moment (cf. Section 4.3) to estimate the parameters; and
- quantiles functions for the three Halphen distributions.

6.2. DSS Code

The first step to make Hydrological Frequency Analysis using the Halphen family distribution, is to determine the most adequate Class (C, D, or E) to represent the dataset. The DSS (cf. Section 3.3 and Figure 3.3) is based on two main approaches:

1) the Log-Log plot (Appendix C.1) and,
2) the Mean Excess Function (Appendix C.2),

with additional two confirmatory approaches,

3) the Hill's report (Appendix C.3) and,
4) the Jackson Statistic (Appendix C.4),

in order to discriminate between the three classes. Maltlab codes are available to run the DSS procedure with input from the vector of the IID dataset where the output corresponds to the most appropriate Class C, D, or E, to represent the distribution tail. The Code **DSS_Main.m** is presented on the next page and gives the final outcome of the DSS. The input corresponds to a vector of Independent and Identically Distributed

(IID) data, and the output is the selected Class C, D, or E. An illustration of the application of this code, is given in Section 6.6.

```
%%%%%%%%%%%%%%%%%%%%%%%%%%%%%%%%%%%%%%%%%%%%%%%%%%%%%%%%%%%%%%%
function [conclusion]=DSS_Main(D)
% The input D : A vector of Idependent and Identically
Distributed Data
% The output : conclusion is the selected Class C, D or E
nbColone=size(D,2);
for k =1:nbColone
    X=D(:,k);
    X = X(find(~isnan (X))); %Delete NaN
    nbLigne(k)= length(X);
    coeficient(k)= abs(DSS_log_log(X));
    Rc5(k)= DSS_EstimationRc5pcent(nbLigne(k));
    S0(k)= abs(DSS_EMEF(X));
    Sc5(k)= DSS_EstimationSc5pcent(nbLigne(k));
    if coefficient(k)>= Rc5(k)
        conclusion(k)= 'C';
    else if S0(k)<Sc5(k)
          conclusion(k) = 'E';
        else
            conclusion(k) = 'D';
        end;
    end
 end

%%%%%%%%%%%%%%%%%%%%%%%%%%%%%%%%%%%%%%%%%%%%%%%%%%%%%%%%%%%%%%%
```

6.3. GENERATION OF HALPHEN DISTRIBUTIONS

Matlab codes are developed to generate samples from Halphen distributions, based on the Accept-Rejection (A-R) approach (cf. Chapter 5). From the input that needs to be generated, three functions, **Gener_HA**, **Gener_HB,** and **Gener_HIB**, are available and have the values of the parameters and the length of the vector. The summary of these functions is presented below.

```
function V = Gener_HA(m,alpha,nu,nlength)
%%%%%%%%%%%%%%%%%%%%%%%%%%%
%%%%%%%%%%%%%%%%%%%%%%%%%%%
%%% Generator of the Halphen type B HB distribution
%%% With parameters nu, alpha and m
%%% Algorithm based on the Accept Reject method
%%% nu should be large than 1 to have finite variance
%%% Developed by: Salah El Adlouni - January 2016
%%%%%%%%%%%%%%%%%%%%%%%%%%%%%%%%%%%%%%%%%%%%%%%%%%%%%%%%%%%%%%%
```

```
%%%%%%%%%%%%%%%%%%%%%%%%%%%%%%%%%%%%%%%%%%%%%%%%%%%%%%%%%%%%
function V= Gener_HB(m,alpha,nu,nlength)
%%%%%%%%%%%%%%%%%%%%%%%%%%
%%% Generator of the Halphen type B HB distribution
%%% With parameters nu, alpha and m
%%% Algorithm based on the Accept Reject method and
%%% The relationship between and the fact that
%%% If X ~ HB (m,alpha,nu) then Y=1./X ~ HIB(1/m,alpha,nu)
%%% nu should be large than 1 to have finite variance
%%% Developed by: Salah El Adlouni - January 2016
%%%%%%%%%%%%%%%%%%%%%%%%%%%%%%%%%%%%%%%%%%%%%%%%%%%%%%%%%%%%

%%%%%%%%%%%%%%%%%%%%%%%%%%%%%%%%%%%%%%%%%%%%%%%%%%%%%%%%%%%%
function V = Gener_HIB(m,alpha,nu,nlength)
%%%%%%%%%%%%%%%%%%%%%%%%%%
%%%%%%%%%%%%%%%%%%%%%%%%%%
%%% Generator of the Halphen type B HB distribution
%%% With parameters nu, alpha and m
%%% Algorithm based on the Accept Reject method
%%% nu should be large than 1 to have finite variance
%%% Developed by: Salah El Adlouni - January 2016
%%%%%%%%%%%%%%%%%%%%%%%%%%%%%%%%%%%%%%%%%%%%%%%%%%%%%%%%%%%%
```

To run these codes write the functions' names for the values of the parameters. For example, to simulate $n=10$ values from $HA(m=100,\alpha=3,\nu=2)$ put the command:

<u>The Input for the HA Distribution</u>

>>V_HA = Gener_HA(100,3,2,10)

<u>The output:</u>

```
L = 2.572872,    U = 766.035358   (Bounds of the instrumental distribution)
maximum = 1.461952                (The constant M of the A-R method)
V_HA = 181.58                     (The simulated vector)
       116.48
       156.6
       118.61
       248.01
       150.81
       183.89
       156.22
       111.65
        76.453
```

The Input for the HB Distribution

To simulate $n = 10$ values from $HB(m = 100, \alpha = 2, \nu = 3)$ input the command:

```
>> V_HB = Gener_HB(100,2,3,10)
```

The output:

```
L = 0.000579, U = 0.009010      (Bounds of the instrumental
                                distribution)
maximum = 3.836179              (The constant M of the A-R method)
V_HB = 98.111                   (The simulated vector)
       149.94
       283.01
       295.8
       245.09
       228.89
       310.91
       112.56
       286.84
       260.26
```

The Input for the HIB Distribution

To simulate $n = 10$ values from $HIB(m = 1000, \alpha = 1, \nu = 3)$ input the command:

```
>> V_HIB = Gener_HIB(1000,1,3,10)
```

The output:

```
L = 67.31,   U = 555.81         (Bounds of the instrumental
                                distribution)
maximum = 2.61                  (The constant M of the A-R method)
V_HB =                          (The simulated vector)
            494.17
            669.79
            893.8
            368.08
            826.77
            515.3
           1667.6
            627.55
            600.37
            410.04
```

6.4. ESTIMATION OF THE PARAMETERS OF THE HALPHEN DISTRIBUTIONS (METHOD OF MOMENTS)

The Matlab codes, to estimate the parameters of the Halphen distributions presented in this section, are based on the Method of Moments (cf. Section 4.3).

HA Distribution

For the Halphen type A distribution, the Matlab function is:

```
%%%%%%%%%%%%%%%%%%%%%%%%%%%%%%%%%%%%%%%%%%%%%%%%%%%%%%%%%%%%
function Param= HA_MM(A)
%%%%%%%%%%%%%%%%%%%%%%%%%%
%%% Parameter estimator of the Halphen type A distribution
%%% For the Vector of dataset A
%%%   Algorithm based on the Method of Moments (MM)
%%%   The output is the vector : Param =[m_est alpha_est
nu_est]
%% Salah El Adlouni %% January 2016
%%%
%%%%%%%%%%%%%%%%%%%%%%%%%%%%%%%%%%%%%%%%%%%%%%%%%%%%%%%%%%%%
%%%%%%%%%%%%%%%%%%%%%%%%%%%%%%%%%%%%%%%%%%%%%%%%%%%%%%%%%%%%
```

For example, to estimate the parameters of the generated vector "V_HA" from $HA(m=100, \alpha=3, \nu=2)$ (cf. Section 6.3), use the command:

>>Param= HA_MM(V_HA)

The output is:

>>Param = [86.581, 4.2204, 4.6294]

HB distribution

For the Halphen type B distribution, the Matlab function is:

```
%%%%%%%%%%%%%%%%%%%%%%%%%%%%%%%%%%%%%%%%%%%%%%%%%%%%%%%%%%%%
function Param_est = HB_MM(A)
%%%%%%%%%%%%%%%%%%%%%%%%%%
%%% Parameter estimator of the Halphen type B distribution
%%% For the Vector of dataset A
%%%   Algorithm based on the Method of Moments (MM)
%%%   The output is the vector: Param =[m_est alpha_est
nu_est]
%% Salah El Adlouni %% January 2016
%%
%%%%%%%%%%%%%%%%%%%%%%%%%%%%%%%%%%%%%%%%%%%%%%%%%%%%%%%%%%%%
```

For example, to estimate the parameters of the generated vector "V_HB" from $HB(m=100, \alpha=2, \nu=3)$ (cf. Section 6.3), use the command:

>>Param= HB_MM(V_HB)
The output is:

>>Param = [240.15 , -3.7568 , 2.7676]

The biggest differences between the theoretical and estimated values are due to the sample size *n*=10, for this illustrative examples.

HIB distribution

For the Halphen type Inverse B distribution the Matlab function is:

```
%%%%%%%%%%%%%%%%%%%%%%%%%%%%%%%%%%%%%%%%%%%%%%%%%%%%%%%%%%%
function Param_est = HIB_MM(A)
%%%%%%%%%%%%%%%%%%%%%%%%%
%%%Parameter estimator of the Halphen type HIB distribution
%%% For the Vector of dataset A
%%%   Algorithm based on the Method of Moments (MM)
%%%   The output is the vector: Param =[m_est alpha_est
nu_est]
%% Salah El Adlouni %% January 2016
%%
%%%%%%%%%%%%%%%%%%%%%%%%%%%%%%%%%%%%%%%%%%%%%%%%%%%%%%%%%%%
```

For example to estimate the parameters of the generated vector "V_HIB" from $HIB(m=1000, \alpha=1, \nu=3)$ (cf. Section 6.3), use the command:

>>Param= HIB_MM(V_HIB)

The output is:

>>Param = [461, -5.2415, 2.7056]

As mentioned in the case of the HB distribution used to simulate the dataset, the estimated values are not close to the theoretical values of the parameters. However, for large sample sizes the estimation method leads to very small bias (cf. Chapter 4).

For example, for the same parameters of the HIB as in the last example, and with large sample size n=1000, the commands to generate the sample and to estimate the parameters are:

>> V= Gener_HIB(1000,1,3,1000);

L = 24.358998, *U = 2481.841402*

maximum = 5.570033

>> Param=HIB_MM(V)

D = 1.0688e + 006

Param = [1033.8, 1.5152, 2.7141]

6.5. QUANTILE FUNCTIONS

The main objective of the hydrological frequency analysis is to estimate the quantile corresponding to a given probability of non-exceedance $q = 1 - p$. Such estimation is essential for risk estimation. For the Halphen distribution, the quantile function does not have an explicit form. The quantile Q_p of the probability of non-exceedance p, is the solution to the equation:

$$\int_0^{Q_p} f_{Halphen}(z; m, \alpha, \nu)\, dz = q$$

When the parameter estimations, $\hat{m}, \hat{\alpha}$, and $\hat{\nu}$, are available, then the solution is the estimated quantile $\hat{Q}_p$:

$$\int_0^{\hat{Q}_p} f_{Halphen}(z; \hat{m}, \hat{\alpha}, \hat{\nu})\, dz = q .$$

To solve this equation, a Matlab code is developed for the three Halphen distributions. The Matlab functions are summarized hereafter.

<u>Quantile function for HA</u>

```
%%%%%%%%%%%%%%%%%%%%%%%%%%%%%%%%%%%%%%%%%%%%%%%%%%%%%%%%%%%
function Qt = HA_quant(q,m,alpha,nu)
%%%%%%%%%%%%%%%%%%%%%%%%%%%%%%%%%%%%%%%%%%%%%%%%%%%%%%%%%
%%% Quantile estimation of the Halphen type A distribution
%%% For the probability of non-exceedance q in [0,1]
```

```
%%% The estimated value of the parameters
%%% The output is the value of the quantile Qt
%% Salah El Adlouni %% January 2016
%%%%%%%%%%%%%%%%%%%%%%%%%%%%%%%%%%%%%%%%%%%%%%%%%%%%%%%%%%%%
```

>> Qt = HA_quant(0.99 , 86.58 , 4.22 , 4.62)

>> Qt = 288.31 (Note that the theoretical value of this quantile is Qth=320)

Note that the bias related to the parameter estimation is due to the sample size, in this case n=10. An important bias for the parameters does not automatically cause a large bias for the quantile estimates. Different combinations of parameters may correspond to distributions with similar statistical characteristics, such as moments and quantiles.

Quantile function for HB

A similar function is available to estimate the quantile of the HB distribution for the given value of the probability and the estimated values of the parameters.

```
%%%%%%%%%%%%%%%%%%%%%%%%%%%%%%%%%%%%%%%%%%%%%%%%%%%%%%%%%%%%
function Qt = HB_quant(q,m,alpha,nu)
%%%%%%%%%%%%%%%%%%%%%%%%%
%%% Quantile estimation of the Halphen type B distribution
%%% For the probability of non-exceedance q in [0,1]
%%% The estimated value of the parameters
%%% The output is the value of the quantile Qt
%% Salah El Adlouni %% January 2016
%%
%%%%%%%%%%%%%%%%%%%%%%%%%%%%%%%%%%%%%%%%%%%%%%%%%%%%%%%%%%%%
```

For example:

Qt = HB_quant(0.99 , 240.15 , -3.7568 , 2.7676)

Qt = 449.8 (Note that the theoretical value of this quantile is

Qth=358.57)

Quantile function for HIB

A similar function is available to estimate the quantile of the HIB distribution for the given value of the probability and the estimated values of the parameters.

```
%%%%%%%%%%%%%%%%%%%%%%%%%%%%%%%%%%%%%%%%%%%%%%%%%%%%%%%%%%%%
function Qt = HIB_quant(q,m,alpha,nu)
%%%%%%%%%%%%%%%%%%%%%%%%%%
%%% Quantile estimation of the Halphen type Inverse B
distribution
%%% For the probability of non-exceedance p in [0,1]
%%% The estimated value of the parameters
%%% The output is the value of the quantile Qt
%% Salah El Adlouni %% January 2016
%%
%%%%%%%%%%%%%%%%%%%%%%%%%%%%%%%%%%%%%%%%%%%%%%%%%%%%%%%%%%%%
```

For example:

For the generated sample from HIB with size $n = 10$:

>> Qt=HIB_quant(0.99,461,-5.24,2.70)

Qt= 1999.2

In the case of sample with size $n = 1000$

>> Qt = HIB_quant (0.99 , 1033.8 , 1.5152 , 2.7141)

Qt = 1252 (Note that the theoretical value of this quantile is Qth=1226.9)

These functions use numerical evaluation of Bessel (Eq 2.3) and Exponential factorial (Eq. 2.29) for integral approximation..

6.6. CASE STUDY

In this section, the annual instantaneous peak flows data of the Potomac River at Point of Rocks for the time period 1895-2006 (water year October-September), are considered to illustrate the use of the HFG codes. Figure 6.1 shows the observed annual peak flow time series. Smith (1985), Katz et al. (2002), and El Adlouni et al. (2008) analyzed the same time series for the time period 1895-1986, 1895-2000, and 1895-2006, respectively. To check whether the observations are independent and the

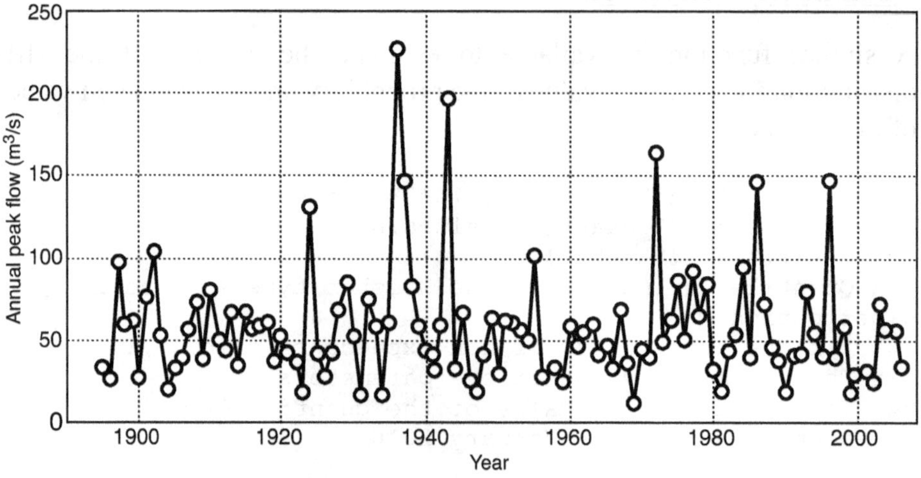

Figure 6.1: Annual peak flows of the Potomac River at Point of Rocks 1895-2006.

data series are stationary and homogeneous we applied the Wald-Wolfowitz, Kendall, and Wilcoxon tests (Bobée and Ashkar, 1991). These tests indicated that the observed peak flow data series can be considered as independent and identically distributed.

The DSS diagram (Figure 3.6) show that the Log-Log plot should be used at first to select the most appropriate class of distribution to represent the observed dataset. The Log-Log plot decision (Figure 6.2) is based on the linearity of the curve. The observed coefficient of correlation $|ro| = 0.88$ is lower than the critical value at the level of 5% ($rc = 0.94$). Given that $|ro| \leq rc$ (non-linearity), then the DSS suggests the use the Mean Excess Function (MEF).

Figure 6.3 presents the MEF diagram for the Potomac peak flow. Note that when the slope is significantly different to zero, at level 5%, the DSS suggests the use of a distribution of the Class D (Figure 3.6). For the Potomac dataset, the slope is $\hat{a} = 0.017$ whereas the critical value, for the same sample size, is $a_0 = 0.004$. Since $\hat{a} > a_0$, then the null hypothesis (H0: The slope is equal to 0), is rejected at the significance level 5%. Results indicate significant positive slope and suggest the use of a distribution of the Class D (Sub-exponential distributions, i.e. Halphen A (HA), Halphen B (HB), Gamma (Gam), Gumbel (EV1), Pearson type 3…; Figure 3.3).

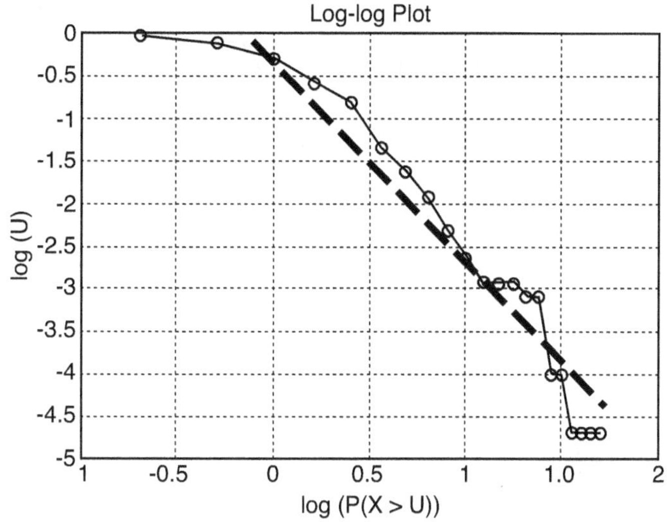

Figure 6.2: Log-Log plot to select the class of heavy tailed distributions for the Potomac River peak flow.

Given that two Halphen distributions HA and HB belong to the same Class D, we suggest the use of the test based on the Delta diagram to discriminate between these distributions and their limiting case Gamma (Section 3.5). Figure 6.4 presents the Delta diagram with the Gamma curve and both limiting bound as developed in El Adlouni et al. (2015).

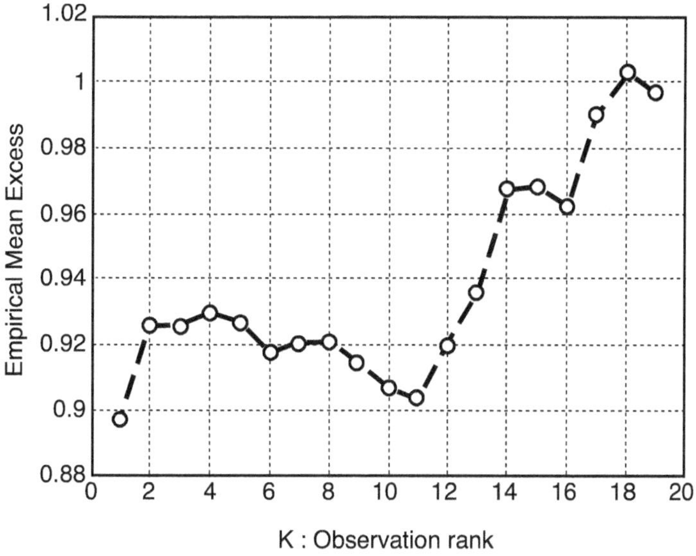

Figure 6.3: Mean Excess Function diagram to discriminate between the Classes D and E for the Potomac River peak flow.

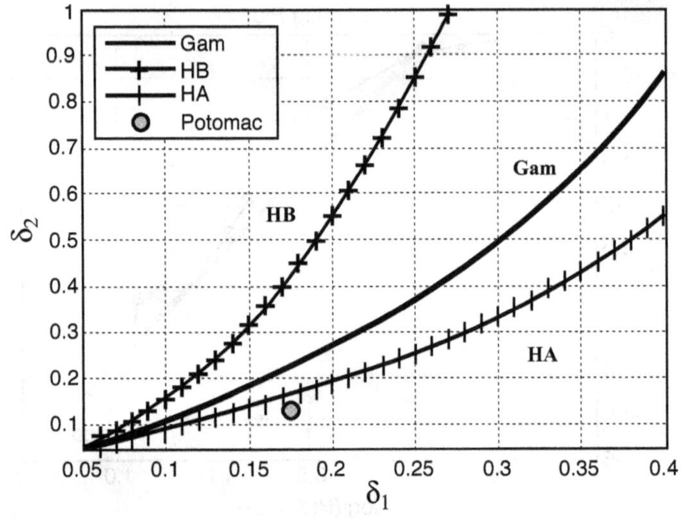

Figure 6.4: The Delta diagram test to discriminate between the HA, HB, and their limiting distribution Gamma.

The position of the point representing the Delta values of the Potomac River dataset in the HA zone, indicates that the HA distribution is the most adequate to fit the data. The estimates of the HA parameters, given by the method of moments (Section 6.4), for the Potomac River dataset are:

$$\upsilon = -1.66; \alpha = 1.37 \text{ and } m = 82.11.$$

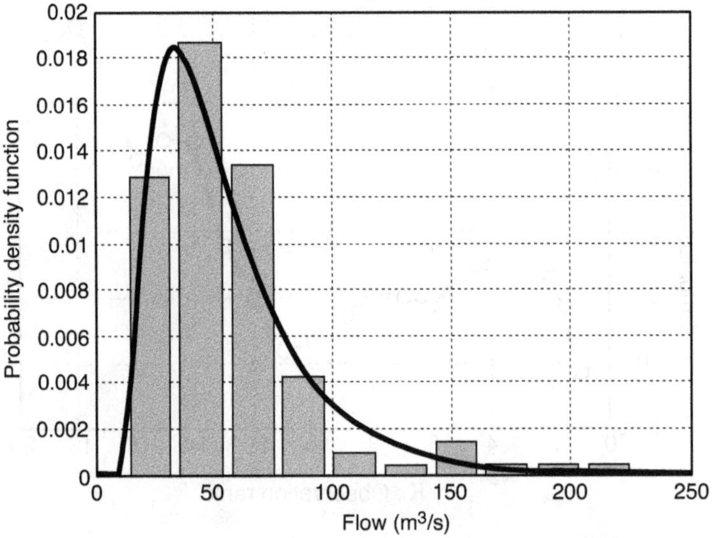

Figure 6.5: The histogram and the probability density function of the HA distribution fit for the Potomac dataset.

6.7. CONCLUSION

The Matlab codes for Halphen Fitting and Generating (HFG) presented in this chapter are user-friendly and can be considered for statistical inference or research study with Halphen family distributions. All the developed functions should be regrouped in the same folder. The sample generation codes of the Halphen datasets for given values of the parameters, could be used for Monte Carlo simulations and bootstrapping, and for the DSS for tail behavior studies. The execution of all the proposed functions is very fast even for large datasets. These tools are more adapted for research studies concerning Halphen distribution, and are complementary to the more general HYFRAN-PLUS software (https://www.wrpllc.com/books/HyfranPlus/indexhyfranplus3.html).

Appendices

APPENDIX A: Bessel Functions

The normalizing constant of the Halphen type A distribution is based on the second modified Bessel function for the argument z and the index v, $K_v(z)$ (Perreault et al. 1997, Watson, 1966, and Abramovitz and Stegun, 1972). This function is defined for $z > 0$ by:

$$K_v(z) = \frac{1}{2}\int_0^\infty x^{v-1}\exp\left[-\frac{z}{2}\left(x+\frac{1}{x}\right)\right]dx \tag{A.1}$$

The expression of the Bessel function used in the HA distribution Eq. (2.3) is deduced from Eq. (A.1) through the transformation $y = xm$:

$$\begin{aligned} K_v(z) &= \frac{1}{2}\int_0^\infty \left(\frac{y}{m}\right)^{v-1}\exp\left[-\frac{z}{2}\left(\frac{y}{m}+\frac{m}{y}\right)\right]\frac{1}{m}dy \\ &= \frac{1}{2m^v}\int_0^\infty y^{v-1}\exp\left[-\frac{z}{2}\left(\frac{y}{m}+\frac{m}{y}\right)\right]dy \end{aligned}$$

Then, for $z = 2\alpha$:

$$K_v(2\alpha) = \frac{1}{2m^v}\int_0^\infty y^{v-1}\exp\left[-\alpha\left(\frac{y}{m}+\frac{m}{y}\right)\right]dy$$

The function $K_v(z)$ is related to the first modified Bessel function $I_v(z)$ by:

$$K_v(z) = \frac{\pi}{2}\frac{1}{\sin(\pi v)}\left[I_{-v}(z) - I_v(z)\right] \tag{A.2}$$

Where $I_v(z)$ is defined by:

$$I_v(z) = \sum_{k=0}^\infty \frac{(z/2)^{2k+v}}{m!\Gamma(k+v+1)} \tag{A.3}$$

A.1. Recursive relations

The Bessel function $K_v(z)$ satisfies the following recursive relation:

$$K_{v-1}(z) - K_{v+1}(z) = \frac{2v}{z} K_v(z) \tag{A.4}$$

$$K_{v-1}(z) - \frac{v}{z} K_v(z) = \frac{\partial K_v(z)}{\partial z} \tag{A.5}$$

$$K_{v-1}(z) + K_{v+1}(z) = 2\frac{\partial K_v(z)}{\partial z} \tag{A.6}$$

$$\frac{v}{z} K_v(z) + K_{v+1}(z) = \frac{\partial K_v(z)}{\partial z} \tag{A.7}$$

$$K_v(z) = K_{-v}(z) \tag{A.8}$$

A.2. Asymptotic properties

From Eqs. (A.2) and (A.3), it is possible to deduce limiting forms of the Bessel function $K_\upsilon(z)$.

- For small value z:

$$K_v(z) \cong \frac{2^{v-1}}{z^v} \Gamma(v), \quad v > 0 \tag{A.9}$$

- For large z:

$$K_v(z) = \sqrt{\frac{\pi}{2}}\, z^{-1/2} e^{-z} \left[\begin{array}{c} 1 + \dfrac{u-1}{8z} + \dfrac{(u-1)(u-9)}{2!(8z)^2} \\ + \\ \dfrac{(u-1)(u-9)(u-25)}{3!(8z)^3} + \mathrm{L} \end{array} \right] \tag{A.10}$$

where, $u = 4v^2$.

- For large value of v:

$$K_v(z) \cong \sqrt{\frac{\pi}{2}} 2^v v^{v-1/2} e^{-v} z^{-v} \tag{A.11}$$

A.3. Dispersion function $D_A(\alpha,v)$

The dispersion function is defined by Eq. (4.5a) $D_A(\alpha,v) = A/H$ and is related to Bessel function (Eq. (2.18)) by:

$$D_A(\alpha,v) = \frac{K_{v+1}(2\alpha)K_{v-1}(2\alpha)}{\left[K_v(2\alpha)\right]^2} .$$

Resolution of the maximum likelihood (ML) system is based on the properties of the dispersion function. The following proposition establish the main properties of the function $D_A(\alpha,v)$.

Proposition A.1. *The dispersion function $D_A(\alpha,v)$, for given v, is a positive decreasing function of α and:*

$$\lim_{\alpha\to 0} D_A(\alpha,v) = \begin{cases} +\infty & \text{si } |v|\le 1 \\ |v|/(|v|-1) & \text{si } |v|>1 \end{cases} \tag{A.12}$$

$$\lim_{\alpha\to +\infty} D_A(\alpha,v) = 1 \tag{A.13}$$

Figure A.1 illustrates the behavior of the dispersion function $D_A(\alpha,v)$ as function of α, for different values of the parameter v.

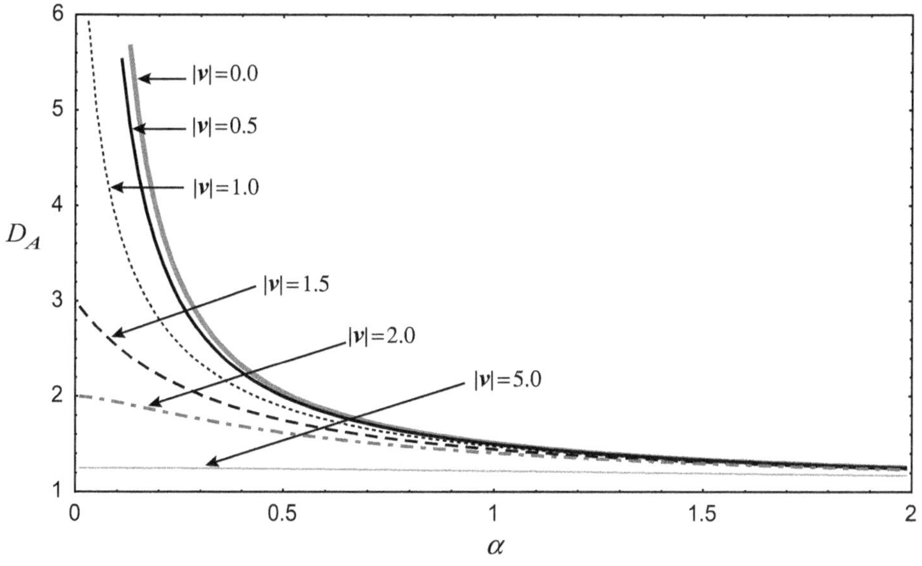

Figure A.1. Dispersion function $D_A(\alpha,v)$.

Proposition A.1 indicates that the maximum likelihood (ML), Eq. (4.5a) doesn't have a solution for all dataset. Indeed, since the dispersion function has values greater than 1, then the system of the ML for the HA distribution doesn't have any solution when the arithmetic mean is equal to the harmonic mean (*A* = *H*). In addition, when $\alpha \to 0$, the dispersion function $D_A(\alpha, v)$ could be bounded for some values of the parameters v (cf. Figure A.1). In this case, the ML estimators of α and m will not correspond to the solution of the system of Eq. (4.5).

For $\alpha \to 0$, Proposition A.1 indicates that the Eq. (4.5a) doesn't have any solution when $|v| \leq 1$ or $AH^{-1} < |v|/(|v|-1)$ for $|v| > 1$, because of the upper bound of the function $D_A(\alpha, v)$. The last inequality is equivalent:

$$|v| > \left(|v| - 1\right) AH^{-1}$$

or

$$|v| < AH^{-1}/(AH^{-1} - 1)$$

Let $U = AH^{-1}/(AH^{-1} - 1)$, then $U > 1$ because $H \leq A$, and when considering the properties of $D_A(\alpha, v)$, the ML estimators of the parameters α and *m*, of the HA distribution, are solutions of the system of Eqs. (4.5a,b).

APPENDIX B: Exponential Factorial Function

The Exponential factorial function for the argument α and indexed by v, noted $ef_v(\alpha)$ is defined by (Halphen, 1955):

$$ef_v(\alpha) = \Gamma(v) + \Gamma(v+1/2)\frac{\alpha}{1!} + \cdots + \Gamma(v+r/2)\frac{\alpha^r}{r!} + \cdots \qquad \text{(B.1)}$$

where $\Gamma(.)$ is the gamma function. The $ef_v(\alpha)$ (Eq. (2.29)) for $v > 0$, can be presented as integral function as:

$$ef_v(\alpha) = 2\int_0^\infty x^{2\upsilon-1} \exp\left[-x^2 + \alpha x\right] dx \qquad \text{(B.2)}$$

The exponential factorial function $ef_v(\alpha)$ is related to some particular functions such as the parabolic cylindrical function $U(\gamma, \beta)$ and confluent hypergeometric function $M(a, b, z)$ defined by (cf. Abramovitz and Stegun, 1972):

$$U(\gamma, \beta) = \frac{e^{-\beta^2/4}}{\Gamma(\gamma+1/2)}\int_0^\infty t^{\gamma-1/2} \exp\left(-\beta t - \frac{t^2}{2}\right) dt \qquad \text{(B.3)}$$

$$M(a, b, z) = \frac{\Gamma(b)}{\Gamma(b-a)\Gamma(a)}\int_0^1 e^{zt} t^{a-1}(1-t)^{b-a-1}\, dt \qquad \text{(B.4)}$$

and show that:

$$ef_v(\alpha) = \frac{\Gamma(2v)e^{\alpha^2/8}}{2^{v-1}} U\left(2v - \frac{1}{2}, \frac{-\alpha}{\sqrt{2}}\right) \qquad \text{(B.5)}$$

$$ef_v(\alpha) = \Gamma(v)\, M\left(v, \frac{1}{2}, \frac{\alpha^2}{4}\right) + \alpha\,\Gamma\left(v + \frac{1}{2}\right) M\left(v + \frac{1}{2}, \frac{3}{2}, \frac{\alpha^2}{4}\right) \qquad \text{(B.6)}$$

These relationships are useful in practice in order to evaluate the exponential factorial function, when routines are available for the other functions.

B.1. Recurrence relations

The exponential factorial function $ef_v(\alpha)$ satisfies the following recurrence relations (Halphen, 1955):

$$ef_{v+1}(\alpha) = \frac{\alpha}{2} ef_{v+1/2}(\alpha) + v\, ef_v(\alpha) \tag{B.7}$$

$$\frac{\partial^n ef_v(\alpha)}{\partial \alpha^n} = ef_{v+n/2}(\alpha) \tag{B.8}$$

B.2. Asymptotic relations

The exponential factorial function $ef_v(\alpha)$ satisfies the following limiting forms (Halphen, 1955):

- For α approaches $-\infty$, then:

$$ef_v(\alpha) \cong 2\Gamma(2v)\frac{1}{|\alpha|^{2v}}\left[1 - \frac{2v(2v+1)}{1!}\frac{1}{\alpha^2} + \cdots\right], \quad v > 0 \tag{B.9}$$

- For α large, we have:

$$ef_v(\alpha) = 2\sqrt{\pi}\left(\frac{\alpha}{2}\right)^{2v-1} e^{\alpha^2/4}\left[1 + (2v-1)(2v-2)\frac{1}{\alpha^2} + \cdots\right] \tag{B.10}$$

B.3. Dispersion function $D_B(\alpha, v)$

The dispersion function is defined in Eq. (2.38) and related to the exponential factorial function by:

$$D_B(\alpha, v) = \frac{ef_{v+1}(\alpha) ef_v(\alpha)}{\left[ef_{v+1/2}(\alpha)\right]^2}$$

The proposition B.1 presents the principal properties of the function $D_B(\alpha, v)$ which are important to resolve the maximum likelihood system for the HB and HIB distributions (cf. Sections 4.2.2 and 4.2.3, respectively).

PROPOSITION B.1. The dispersion function $D_B(\alpha,v)$ *is, for fixed value of* υ, *positive and decreasing for* α *and:*

$$\lim_{\alpha\to-\infty} D_B\left(\alpha,v\right)=1+\frac{1}{2v} \qquad \text{and} \qquad \lim_{\alpha\to+\infty} D_B\left(\alpha,v\right)=1$$

For the demonstration of this proposition see Perreault et al. (1997).

Figure B.1 illustrates the behavior of the function $D_B(\alpha,v)$ as function of α, for different values of the parameter v.

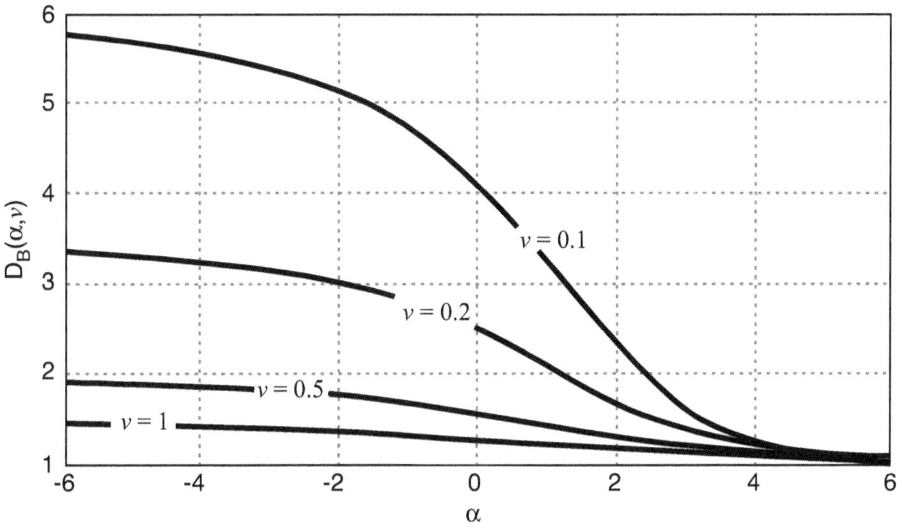

Figure B.1. The dispersion function $D_B(\alpha,v)$**.**

The proposition in Figure B.1 shows that the Eq. (4.9a) does not admit always a solution. Indeed, $D_B(\alpha,v)$ is all time greater than 1, then the system of Eqs. (4.9a,b) has not a solution when $Q=A^2$, i.e. in the case of identical observations, which is impossible in practice. In addition, when $\alpha\to-\infty$, the function $D_B(\alpha,v)$ is bounded whatever the value of v. In this case, the maximum likelihood estimators correspond to the solution of the system of Eqs. (4.9a,b), for fixed value of υ, if and only if, the observations are not identical and $Q/A^2<1+(1/2v)$. This last inequality indicates that the dispersion of the dataset is bounded.

The inequality $Q/A^2 < 1 + (1/2v)$ is equivalent to:

$$\frac{1}{v} > 2\left(\frac{Q}{A^2} - 1\right)$$

or

$$v < \frac{1}{2}\left(\frac{Q}{A^2} - 1\right)^{-1}$$

Let $V = 1/[2(QA^{-2} - 1)]$, then $0 < V < \infty$ because $Q \geq A^2$, and given the properties of $D_B(\alpha, v)$, the maximum likelihood estimators of the parameters α and m are solutions of the system of Eqs. (4.9a,b), if and only if $0 < v < V$. The statistic V, as in the case of the HA distribution with the statistic U, defines for the HB distribution the interval of the values of the parameter v that ensure the existence of the solution of the system of Eqs. (4.9a,b).

When the condition $0 < v < V$ is not satisfied, i.e. if $Q / A^2 \geq 1 + (1/2v)$, then the dispersion function is large and the maximum likelihood estimators of α and m converge to that of the Gamma distribution. In this case, i.e. $v \geq V$, the Gamma distribution is the most adequate than the HB distribution, to model the observed dataset. Indeed, for a given value of v, the limiting Gamma distribution has the largest coefficient of variation among all HB distributions (cf. Section 2.3.2).

APPENDIX C: Decision Support System

In the four following section, we present detail on the criteria considered in the Decision Support System (DSS) as illustrated in the diagram (Figure 3.6). The main approaches are the Log-Log plot and Mean Excess Function (MEF) in order to discriminate between the Classes C, D, and E. Two other methods are considered for confirmatory analysis.

C.1. Log-Log Plot

The Log-Log plot (Beirlant et al. 2006) is based on the fact that the survival function $\bar{F}(u) = P(X > u)$, is given by

$$\bar{F}(u) = P(X > u) = \exp\{-u/\theta\}$$

for exponential tail with mean θ; for regularly varying distribution with tail index α, $\bar{F}$ is equivalent, for large quantile, to:

$$\bar{F}(u) = P(X > u) \approx C\int_u^\infty \frac{1}{x^\alpha}dx = C\left[\frac{x^{-\alpha+1}}{1-\alpha}\right]_u^\infty = C_1 u^{-\alpha+1}$$

(with $\alpha > 1$), which is equivalent to finite mean).

Therefore, using the logarithmic transformation, the probability to verify regularly varying distributions, is

$$\log[P(X > u)] \approx \log C_1 - (\alpha - 1)\log(u).$$

This suggests that, in the Log-Log plot ($\log[P(X > u)]$ vs $\log(u)$), the tail probability is represented by a straight line for power-law (or regularly varying distributions, Class C), but not for the other sub-exponential or exponential distributions (Class D or E).

Thus, the curve represented in the Log-Log plot will correspond to a straight line for the distributions of the Class C i.e. Fréchet (EV2), Halphen type IB (HIB), Log-Pearson type 3 (LP3) and Inverse Gamma (IG) (cf. Figure 3.3), but not for sub-exponential or exponential type tails (Class D or E). When the diagram is not linear we suggest (cf. Figure 3.6) using the Mean Excess Function (MEF) to discriminate between the Classes D and E (cf. Section C.2).

To check the linearity of the curve in the log-log diagram, a test on the associated coefficient of correlation is considered. Simulation studies allow the determination of critical values (rc) corresponding to significance levels of 5% and 1%, to test the hypothesis H0: the data follow a distribution of the Class C (i.e. the curve is linear). The hypothesis H0 is equivalent to H0′: theoretical value of the coefficient of correlation $\rho = 1$. These critical values are calculated according to the size N of the sample ($30 \leq N \leq 200$). Note that the decisions given by the DSS are based, by default, on the significance level 5 %.

If the hypothesis H0 is rejected, at the significance level 5%, we suggest the use of the mean excess function plot (MEF). Indeed, if the observed correlation coefficient (ro) is greater than critical value (rc) at the significance level 5%, then we conclude that it is not significantly different from 1 at the significance level 5 % and the hypothesis H0 of linearity is accepted at this level. In this case, the most adequate choice corresponds to the Class C of regularly varying distributions (power-law type): HIB, EV2, LP3, IG.

C.2. The Mean Excess Function Diagram (MEF)

The mean excess function method (Beirlant et al., 2006) is based on the function $e(u) = E[X - u \mid X > u]$. This function is constant for exponential tail distributions ($e(u) = \theta$). However, in the case of regularly varying distribution with tail index α ($\alpha > 2$): $e(u) = (u / (\alpha - 2))$. The Mean Excess Function (MEF) allows discriminating between the Class D (sub-exponential distributions) and the Class E (Exponential distribution). Indeed, the curve presented in the MEF diagram is linear for high observed values for distributions of both Classes D and E. If in addition the slope of this curve is:

- Null, the most adequate distribution belongs to the Class E (Exponential distribution);
- Strictly positive, the most adequate distribution belongs to the Class D of sub-exponential distributions i.e. HA, EV1, HB, P3, and G (cf. Figure 3.3).

Note that, in the DSS, this method should be used after the Log-Log plot method. Indeed, if the assumption H0 of the Log-Log plot method is rejected (distribution does not belong to the Class C of regularly varying distributions and then it belongs to Class D) FME method allows testing whether the distribution is exponential or not.

The use of this diagram in the DSS is based on the slope of the MEF curve for the observations that exceed the median (50 % of the highest observed value of the sample because it is an asymptotic test).

Simulation studies allow the determination of critical values of the slope corresponding to significance levels of 5 % and 1 %, to test the hypothesis H0: the data follow a distribution of the Class E (i.e. the slope of the MEF is equal to zero). These critical values are calculated according to the size N of the sample ($30 \leq N \leq 200$). Note that the decisions given by the DSS are based, by default, on the significance level 5 %. When the hypothesis H0 is accepted we suggest the use of the Exponential distribution (Class E). However, when it is rejected at the significance level 5 %, we suggest the use of a distribution of the Class D (HA, EV1, HB, P3, G).

C.3. Hill's Ratio Plot

The Hill ratio (Hill, 1975) is defined by

$$a_n(x_n) = \frac{\sum_{i=1}^{n} \mathrm{I}(X_i > x_n)}{\sum_{i=1}^{n} \log(X_i / x_n)\mathrm{I}(X_i > x_n)}$$

where

$$\mathrm{I}(X_i > x_n) = \begin{cases} 1 & \text{if} \quad X_i > x_n \\ 0 & \text{if} \quad X_i < x_n \end{cases}.$$

This method is based on the fact that α_n is a consistent estimator of α if the tail is regularly varying (Class C) with tail index α. In the expression of the Hill ratio, x_n is chosen to be large such that $P(X > x_n) \to 0$ and $nP(X > x_n) \to \infty$, and I is the indicator function. The standard Hill estimator, of the tail index, corresponds to the particular case where the observations are ordered $X_{(1)} \leq \ldots \leq X_{(n)}$ and $x_n = X_{(k_n+1)}$, where k_n is an integer which tends to infinity as n tends to infinity.

In practice, the values of the function $\alpha_n(x_n)$ are plotted as function of x_n and the user is looking for some stable region from which $\alpha_n(x_n)$ can be considered as an estimator of α.

This statistics is used in the DSS to confirm the suggested choice given by the first two diagrams (the distribution belongs to the Class C, D, or E).

- If the curve converges to a non-null constant value, the most adequate distribution belongs to the Class C (regularly varying distribution). We suggest then the use of a distribution of the

Class C i.e. Fréchet (EV2), Halphen Inverse type B (HIB), Log-Pearson type 3 (LP3), Inverse Gamma (IG) (cf. Figure 3.3).

- If the curve decreases to zero, the distribution belong to the Sub-exponential class (Class D, i.e. Halphen type A, Gamma, Pearson type III, Halphen type B, Gumbel; and the Exponential class (Class E: Exponential distribution, cf. Figure 3.3).

C.4. Jackson Statistic

This method is presented by Beirlant et al. (2006) and is based on the Jackson statistic (Jackson, 1967). It allows to test whether the sample is consistent with Pareto type distributions. Note that the distributions of the Class C (regularly varying distribution) have asymptotically the same behavior as that of the Generalized Pareto distribution. Originally the Jackson statistic was proposed as a goodness-of-fit statistic for testing exponential behavior, and given the link between the Exponential and the Pareto distribution (if X has a Pareto distribution the logarithmic transformation $Y = \log(X)$ is exponentially distributed) this statistic is used to assess Pareto-type behavior. The Jackson statistic is further modified by taking into account the second-order tail behavior of a Pareto-type model. Beirlant et al. (2006) give the limiting distribution of this statistic with corrected bias version for finite size samples. This modified version of the Jackson statistic converges to 2 for power tail type distribution (Class C) and has an irregular behavior for sub-exponential or exponential distributions (Class D or E). In the DSS, the Jackson statistic is used to characterize distributions of the Class C. Indeed, regularly varying distributions (Class C) has asymptotically a power tail.

In the DSS, this method is considered as a confirmatory method for suggested decision based on the Log-Log and the MEF. So:

- If the curve converges clearly and regularly to 2, the studied distribution belongs to the Class C (regularly varying distribution), i e. Fréchet (EV2), Halphen type IB (HIB), Log-Pearson type 3 (LP3), Inverse Gamma (IG) (cf. Figure 3.3);

- If the curve presents some irregularities and do not converge to 2, than we suggest the sub-exponential Class D, i.e. Halphen type A, Gamma, Pearson type III, Halphen type B, Gumbel; or Class E, exponential (cf. Figure 3.3).

APPENDIX D: Bayesian Framework

The bayesian framework facilitates representing and taking in account the uncertainties related to models and parameter values. In contrast, most decision analyses based on frequency estimation methods involve fixing the values of parameters that may have an important effect on the final outcome of the analysis and for which there is considerable uncertainty. One of the major benefits of the bayesian approach is the ability to incorporate prior information by specifying levels or ranges of individual parameters for use in sensitivity analysis (Bernier, 1967; Parent et al. 2012). The uncertainty on the estimates is presented by the full posterior distribution and can be summarized by credible regions which reflect our degree of belief. In addition, in the bayesian approach the analyst is forced to look at historical data sets or to elicit expert knowledge to determine what is known about the parameters and the processes (Lecoutre, 2005).

The bayesian inference is carried out in the following way:

1. Select a probability distribution $\pi(\theta)$ for the vector of the parameters; called the prior distribution; which expresses the prior beliefs about $\underline{\theta} = (m, \alpha, v)$, before observing the data;

2. Given the model, HA, HB or HIB, and after observing data $(X_1, ..., X_n)$, the posterior distribution $\pi(\theta | X_1, ..., X_n)$ is given through the Bayes rule:

$$\pi\left(\theta \middle| X_1, \ldots, X_n\right) = \frac{f\left(X_1, \ldots, X_n \middle| \theta\right) \pi\left(\theta\right)}{\pi\left(X_1, \ldots, X_n\right)} = \frac{\pi\left(\theta\right) \prod_{i=1}^{n} f\left(X_i \middle| \theta\right)}{\pi\left(X_1, \ldots, X_n\right)} \quad \text{(D.1)}$$

From the bayesian Decision Theory, the bayesian estimators are solutions to an optimization problem of the cost function. For example in the case of a squared cost (SC), the bayesian estimator corresponds to the mean of the posterior distribution and is given by:

$$\tilde{\theta}_{SC} = \int \theta \, \pi\left(\theta \middle| X_1, \ldots, X_n\right) d\theta$$

In the case of 0–1 cost, the bayesian estimator is the mode of the posterior distribution and is given by:

$$\tilde{\theta}_{01} = \arg\max_{\theta} \pi\left(\theta \middle| X_1,\ldots,X_n\right) = \arg\max_{\theta} \log\pi\left(\theta \middle| X_1,\ldots,X_n\right)$$

$$= \arg\max_{\theta} \sum_{i=1}^{n} \log\pi\left(X_i \middle| \theta\right) + \log\pi\left(\theta\right)$$

Note that the normalizing term, corresponding to marginal likelihood can be ignored. However, this is not the case for the first one, where the mean should be computed, usually trough Monte Carlo simulations.

In bayesian statistics a prior distribution is multiplied by the likelihood function and then normalized to produce a posterior distribution (Eq. (4.34)). Some approaches allow to define the prior distributions, such as the non-informative Jeffreys, where no additional information is available on the parameters of the model; or by elicitation to formulate the priors.

A more practical way to define the priors uses the notion of the conjugate distribution family. The conjugate prior is one which, when combined with the likelihood and normalization constant, produces a posterior distribution from the same family as the prior. In most cases, the normalization follows from the form of the distribution family.

As mentioned in Section 2.2.3, the exponential family has interesting properties to ensure the optimality of the maximum likelihood estimators. In the bayesian framework, the main property is the availability of a conjugate prior distribution for the vector of the parameters.

In the case of natural exponential family, where the probability density function of the variable X given the vector of the parameters $\underline{\theta}$, has the form:

$$f\left(x \middle| \theta\right) = \exp\left(s\left(x\right) \bullet \theta + \ln\left(h\left(x\right)\right) - g\left(\theta\right)\right) \qquad \text{(D.2)}$$

where θ represents the vector of the parameters, $g(.)$ is the log-cumulate function that normalizes the pdf, $s(.)$ are the sufficient statistics, and $h(.)$ is a function of the variable alone.

The conjugate prior distribution for θ is given by:

$$\pi\left(\theta\right) \propto \exp\left(\lambda\theta - \alpha g\left(\theta\right)\right) \qquad \text{(D.3)}$$

A prior is defined to be conjugate if the corresponding posterior is of the same form as the prior. Then, the corresponding posterior distribution is given by:

$$\pi\left(\theta|x\right) \propto f\left(x|\theta\right)\pi\left(\theta\right) \propto \exp\left(\left(x+\lambda\right)\theta-\left(1+\alpha\right)g\left(\theta\right)\right) \qquad \text{(D.4)}$$

and it is indeed of the same form as the prior given in Eq. (D.3). The main motivation for using such priors is that the posterior distributions can be derived analytically. The hyperparameters control the strength of the prior. Values of l and α near zero correspond to weakly informative prior.

References

Abramovitz, M. and I. Stegun, (1972), Handbook of Mathematical Functions. With Formulas, Graphs, and Mathematical Tables. Dover Publications, New York, 1060 pages.

Ahrens, J. H. and Dieter, U. (1974). Computer methods for sampling from gamma, beta, Poisson and binomial distributions. Computing12: pp. 223-246.

Alexander, G. N., Morlat, G., O'Connell, P. E., O'Donnell, T., Sneyers, R., Delaporte, P. J., Elston, R. D. and Borgman, L. E. (1970). Heavy Rainfall as a Stochastic Process. Revue de l'Institut International de Statistique / Review of the International Statistical Institute, Vol. 38, No. 1, pp. 62-81.

Andrieu, C. and Roberts, G. (2009). The pseudo-marginal approach for efficient Monte Carlo computations. The Annals of Statistics. 37, pp. 697–725.

Beirlant, J., de Wet, T. and Goegebeur, Y., (2006). A goodness-of-fit statistic for Pareto-type behavior. *Journal of Computational and Applied Mathematics*, 186, pp. 99-116.

Bendjoudi H. and Hubert, P. (1998). A propos de la distribution statistique des cumuls pluviométriques annuels. Faut-il en finir avec la normalité. Revue des Sciences de l'Eau. 4(98), pp. 617-630.

Berman, S. M. (1962). Limiting distribution of the maximum term in sequences of dependent random variables. Annals of Mathematical Statistics, 33 (3), pp. 894-908.

Bernier, J. (1956).Sur l'application des diverses lois limites des valeurs extrêmes au problème des débits de crues. La Houille Blanche (5). Novembre 1956.

Bernier, J. (1959). Comparaison des lois de Gumbel et de Fréchet sur l'estimation des débits maxima de crue. La Houille Blanche (SHF) (1). Janv.-Fév. pp. 47-56.

Bernier, J. (1967). Les méthodes bayésiennes en hydrologie statistique (Essai de reconciliation de l'hydrologue et du statisticien). First International Hydrology Symposium Fort Collins, pp. 461-470.

Bernier, J. (1998). Information, modèles, risques et hydrologie statistique. Comptes-Rendus de la conférence Internationale « Méthodes Statistiques et Approches Bayésiennes en Hydrologie » en l'honneur du professeur Jacques Bernier, Paris, 11-13 septembre 1995. Edited by E. Parent, P. Hubert, B. Bobée and, J. Miquel. IHP-V, UNESCO-Paris / Documents Techniques en Hydrologie, N°20 pp. 23-38.

Bickel, P. J. and K. A. Doksum (1977). Mathematical Statistics. Holden-Day, Inc., California. 340 pages.

Bobée, B. (1999) Extreme flood events valuation using frequency analysis: a critical review. La Houille Blanche, 54 (7-8), pp. 100-105.

Bobée, B. and Ashkar, F. (1988) Generalized Method of Moments Applied to LP3 Distribution. Journal of Hydraulic Engineering (ASCE), 114, pp. 899-909.

Bobée, B. and Ashkar, F., (1991): The Gamma Family and Derived Distributions Applied in Hydrology. Water Resources Publications, Littleton, CO., 203 pages.

Bobée. B., Ashkar, F. and Perreault, L. (1993). Two kinds of moment ratio diagrams and their applications in hydrology. Stochastic Hydrology and Hydraulics, 7, pp. 41-65.

Bobée B., and Robitaille R., (1975). Correction of bias in the estimation of the coefficient of skewness. Water Resources Research. Vol 11. No. 6, pp. 851-854.

Champernowne D. G. (1953). A model of income distribution. Economical Journal, (23), pp. 318-351.

Chebana, F.; El Adlouni, S. and Bobée B. (2008). Method of moments of the Halphen distribution parameters. Stochastic Environmental Research and Risk Assessment. 22 (6) pp. 749-757.

Chebana, F., El Adlouni, S. and Bobée B. (2010). Mixed estimation methods for Halphen distributions with applications in extreme hydrologic events. Stochastic Environmental Research and Risk Assessment.24: pp. 359-376.

CHS-Chaire en Hydrologie Statistique (2002). HYFRAN : Logiciel pour l'analyse fréquentielle en hydrologie. Water Resources Publications, LLC. (http://wrpllc.com/books/hyfran.html).

Devroye L. (1986). Non-Uniform Random Variate Generation, Springer Verlag, New York. 485 pages.

Dvorak, V.; Bobée, B.; Boucher, S. and Ashkar, F. (1988) Halphen distributions and related systems of frequency functions Research Report No. R-236 INRS-Eau. Ste-Foy. Qc. Canada. 81 pages.

El Adlouni (2002). Estimation bayésienne des paramètres de la loi Halphen type A. Thèse de doctorat. Département de mathématiques, Université Mohamed V, Rabat. 141 pages.

El Adlouni, S. and Bobée, B. (2007) Sampling techniques for Halphen distributions. Journal of Hydrologic Engineering. 12, Issue 6, pp. 592-604.

El Adlouni, S. and Bobée, B. (2010). Système d'aide à la décision pour l'estimation du risque hydrologique. Special issue of Journal des Sciences Hydrologiques, presented to FRIEND 2010 conference, Fez, Morocco.

El Adlouni, S., Bobée, B, and Ouarda, T. B.M.J (2008). On the tails of extreme event distributions in Hydrology. Journal of Hydrology, 355, pp. 16-33.

El Adlouni, S., Favre, A.C. and Bobée, B. (2006). Comparison of methodologies to assess the convergence of Markov Chain Monte Carlo methods. Computational Statistics and Data Analysis 50(10): pp. 2685-2701.

El Adlouni S., Hammami D. and Bobée B. (2015). Discrimination test between the Halphen (A and B) and the gamma distributions. Stochastic Environmental Research and Risk Assessment, Vol 29, Issue 1, pp 13-26.

Embrechts, P. (1983). A Property of the Generalized Inverse Gaussian Distribution with Some Applications. Journal of Applied Probability, 20 (3), pp. 537-544.

Embrechts, P., Kleppelberg, C. and Mikosch, T. (2003). Modelling Extremal Events for insurance and Finance, Applications of Mathematics. Volume 33. Springer, 648 pages.

Fisher, R. A. (1922). On the dominance ratio. Proc. Roy. Soc. Edinburgh, 42, pp. 321-341.

Fisher, R. A. and Tippett, L.H.C. (1928) Limiting forms of the frequency distribution of the largest and smallest member of a sample, Proc. Cambridge Phil. Soc., 24, pp. 180-190.

Fitzgerald, D. L. (2000) Satistical aspects of Tricomiís function and modified Bessel functions of the second kind. Stochastic Environmental Research and Risk Assessmen,t 14, pp. 139-158.

Gelfand, A. E. and Smith, A. F. M. (1990). Sampling-based approaches to calculating marginal densities. Journal of the American Statistical Association,. 85, pp. 398-409.

Geman S. and D. Geman (1984). Stochastic relaxation, Gibbs distributions, and the Bayesian restoration of images, IEEE Trans. Pattern Anal. Mach. Intell, 6, pp. 721-741.

Gibrat, R. (1931). Les inégalités économiques. Thèse en Droit de l'Université de Lyon, soutenue le 28 Janvier, Paris, Sirey, 296 pages.

Gumbel, E. J. (1958). The Statistics of Extremes. New York: Columbia University Press, 371 pages.

Halphen, E. (1941) Sur un nouveau type de courbe de fréquence. Comptes Rendus de l'Académie des Sciences, Tome 213, pp. 633-635. Due to war constraints, published under the name ``Dugué".

Halphen, E. (1955) Les fonctions factorielles. Publications de l'Institut de Statistique de l'UniversitÈ de Paris, Vol. IV, Fascicule I, pp. 21-39.

Hastings W. K. (1970). Monte Carlo sampling methods using Markov chains and their applications. Biometrika, Vol. 57, No. 1. (Apr., 1970), pp. 97-109.

Hill, B. M. (1975). "A simple general approach to inference about the tail of a distribution." Annals of Statistics 3, pp. 1163–1174.

Jackson, O.A.Y. (1967). An analysis of departures from the exponential distribution. Journal of the Royal Statistical Society B, 29, pp. 540-549.

Jenkinson, A.F. (1955). The frequency distribution of the annual maximum (or minimum) values of meteorological elements. Q. J. Roy. Meteor. Soc. 87, pp. 158-171.

Jørgensen, B. (1982). Statistical properties of the generalized inverse gaussian distribution. Lecture Notes in Statistics, No 9, Springler-Verlag: 188 pages.

Katz, R. W., Parlange M. B. and Naveau P., (2002). Statistics of extremes in hydrology. *Advances in Water Resources*, 25: pp. 1287-1304.

Kendall, M. and Stuart, A. (1979). The advanced theory of statistics, volume 2. Griffin, London, 4th edition. 474 pages.

Labaye G. (1956). Le problème des évaluateurs de crues de Serre-Ponçon. Essai de détermination d'un optimum économique. Revue de statistique appliquée 4(3), pp. 47-66.

Le Cam L. and G. Morlat (1949). Les lois des débits des rivières francaises. Houille Blanche, Numéro Special B pp. 733–740.

Lecoutre B. (2005). Et si vous étiez un bayésien qui s'ignore ? Revue MODULAD, numéro 32, pp. 92-105.

Mandelbrot, B. (1997), Fractals and Scaling in Finance: Discontinuity, Concentration, Risk, New York: Springer Verlag. 525 pages.

Mandelbrot, B. (2003) Heavy tails in finance for independent or multifractal price increments. Handbook on Heavy Tailed Distributions in Finance, pp. 1-34.

Martins, E.S. and Stedinger, J.R. (2000) Generalized maximum-likelihood generalized extreme-value quantile estimators for hydrologic data. Water Resources Research, 36 (3), pp. 737-744.

Metropolis, N., Rosenbluth, A., Rosenbluth, M., Teller, A. and Teller, E. (1953). Equations of state calculations by fast computing machines. J. Chem. Phys. 21 pp. 1087–1092.

Mitzenmacher M. (2004). A Brief History of Generative Models for Power Law and Lognormal Distributions. Internet Mathematics 1:2, pp. 226–251.

Morlat, G. (1956) Les lois de probabilité de Halphen. Revue de Statistique Appliquée, 3, pp. 21-43.

Neyman, J. (1935). Statistical problems in agricultural experimentation (with discussion). Suppl. J. Roy. Statist. Soc. Ser. B 2(2), pp. 107-180.

Ouarda, T. B.M.J., Ashkar, F., Bensaid, E. and I. Hourani (1994). Statistical distributions used in hydrology, Transformations and asymptotic properties, Scientific Report, Department of Mathematics, University of Moncton, 31 pages.

Parent E., J. Bernier and L. Perreault (2012). Vers l'Inférence bayésienne des lois de Halphen. 44èmes Journées de Statistique: 21-25 Mai 2012, Université Libre de Bruxelles.

Parent E., J. Bernier and L. Perreault (2016). M C et M C M C même Combat Bayesien pour les Lois d.Halphen. Personal Communication.

Perreault, L., Bobée, B. and Rasmussen, P.F. (1997). Les lois de Halphen. Rapport de recherche ~ R-498, Institut National de la Recherche Scientifique, INRS-EAU. 147 pages.

Perreault, L., Bobée, B. and Rasmussen, P.F. (1999a). Halphen Distribution System. I: Mathematical and Statistical Properties. Journal of Hydrologic Engineering, 4, pp. 189- 199.

Perreault, L., Bobée, B. and Rasmussen, P.F. (1999b). Halphen Distribution System. II: Parameter and quantile Estimation. Journal of Hydrologic Engineering, 4, pp. 200- 208.

Reed, W. J. (2002). *On the rank-size distribution for human settlements.* J. Regional Science, 41: pp. 1-17

Reed, W. J. and B. D. Hughes (2002). *On the size distribution of live genera.* Journal of Theoritical Biology, 217 (1) pp. 125-135.

Rice, J. A. (2007). Mathematical statistics and data analysis. 3rd edition. Duxbury Press. 646 pages.

Robert, C.P. (2006), Le Choix Bayésien : Principes et implémentation Springer-Verlag, Paris, 638 pages.

Robson A.J. and Reed W. J. (1999). Flood Estimation handbook, V3: Statistical procedures for flood frequency estimation. Institute of hydrology wilingford. UK. 338 pages.

Rubinstein R.Y. (1981). Simulation and the Monte Carlo Method. John Wiley & Sons. 338 pages.

Seshadri, V. (1993) The Inverse Gaussian Distribution. Clarendon Press, Oxford, UK. 256 pages.

Sichel, H. S. (1971). On a family of discrete distributions particularly suited to represent long-tailed frequency data. In Proceedings of the Third Symposium on Mathematical Statistics (N. F. Laubscher, ed.), S.A. C.S.I.R., Pretoria, pp. 51-97.

Smith, R. L. (1985) Maximum likelihood estimation in a class of non-regular cases, Biometrika, 72, pp. 67-90.

Turcotte, D. L. (1997). Fractals and Chaos in Geology and Geophysics, 2nd ed., Cambridge: Cambridge University Press. 385 pages.

Von Mises, R. (1954). La distribution de la plus grande de n valeurs. In Selected Papers, American Mathematical Society, Providence, RI. Volume II, pp. 271-294.

Watson, G. N. A (1966). Treatise on the Theory of Bessel Functions, 2nd ed. Cambridge, England: Cambridge University Press. 804 pages.

Werner, T. and C. Upper (2002). Time Variation in the tail behavior of Bund Futures Returns. Working paper N∞199. European Central Bank. 36 pages.

Author Index

Subject Index

F

G

H

I

J

K

L

M

N

O

P

R

S

T

V

W

Abbreviations and Symbols

Symbol or Abbreviation	Corresponding notion Alphabetic order	First citation page
$e(\cdot)$	A-R envelope function	79
M	A-R scaling factor	79
A-R	Acceptance-Rejection (A-R)	79
α	Alpha: parameter (Gamma and Halphen)	5
Argmax	Argument of the maximum	118
A	Arithmetic mean	11
$K_v(.)$	Bessel function	8
φ	Characteristic function	9
E	Class of Exponential tail behavior	37
C	Class of regularly varying distributions	37
D	Class of Sub-Exponential distributions	37
C_k	Coefficient of kurtosis (μ_4 / μ_2^2)	12
C_s	Coefficient of skewness ($\mu_3 / \mu_2^{3/2}$)	12
C_V	Coefficient of Variation ($\sqrt{\mu_2} / \mu_1'$)	12
$L(v \mid \alpha, m)$	Conditional likelihood function	57
DSS	Decision Support System	3
Λ	Density function of EV1	37
Φ_ξ	Density function of EV2	38
Ψ_ξ	Density function of EV3	38

l'	Derivative of the log-likelihood function	61
$\psi(\bullet)$	Digamma function	51
$D_A(.,.)$	Dispersion function for HA	13
$D_B(.,.)$	Dispersion function for HB and HIB	21
F_D	Distribution function of D	38
$E[.]$	Expectation of a random variable	11
$ef_v(.)$	Exponential factorial function	17
EVT	Extreme Value Theory	35
EV2	Fréchet distribution	38
Gam	Gamma distribution	16
$\Gamma(\bullet)$	Gamma Function	16
GEV	Generalized Extreme Value distribution	38
GIG	Generalized Inverse Gaussian distribution	16
GML	Generalized Maximum Likelihood	49
G	Geometric mean	11
EV1	Gumbel distribution (or GUM)	37
HFG	Halphen Fitting and Generation	46
HA	Halphen type A distribution	8
HB	Halphen type B distribution	17
HIB	Halphen type Inverse B distributed	24
H	Harmonic mean	11
IS	Importance Sampling	78
IID	Independent and Identically distributed	15
IG	Inverse Gamma distribution	16

IQ	Inverse Quadratic moment	31
IMM	Iterative Method of Moments	58
L	Likelihood function	57
$LogL$	Log-Likelihood function	59
Log-log	Log-Log plot	46
MCMC	Markov Chain Monte Carlo	73
MDA	Max Domain of Attraction	38
ML	Maximum Likelihood	57
MEF	Mean Excess Function	46
MM	Method of Moments	50
MH	Metropolis-Hastings	88
U	ML bound for the HA	61
V	ML bound for the HB	63
W	ML bound for the HIB	65
$R_A(.,.)$	Moment ratio for HA	12
$R_B(.,.)$	Moment ratio for HB	21
δ_1	Moment ratio In(A/G)	6
δ_2	Moment ratio In(G/H)	6
MC	Monte Carlo	78
$\underline{\theta}$	Parameters to be estimated	1
$\pi(\bullet \mid \underline{x})$	Posterior distribution	75
$\pi(\bullet)$	Prior distribution	73
f_D	Probability density function (pdf) of distribution D	6
p	Probability of exceedance	1
q	Probability of non-exceedance	38

$L(v\|\alpha,m)$	Profile likelihood	62
Q	Quadratic moment	24
x_T	Quantile of return period *T* years	1
T	Return period	1
EV3	Reverse Weibull distribution	38
n	Sample size	53
T_i	Sufficient statistics	14
$Supp(\bullet)$	Support of a distribution	79
$\bar{F}_D$	Survival distribution function of *D*	113
$\mu_r(D)$	The *r*th central moment of the distribution D	11
$\mu_D'^{(r)}$	The *r*th non-central moment of the distribution *D*	11
$\psi'(\bullet)$	Trigamma function	52
Var [.]	Variance of random variable	13

www.ingramcontent.com/pod-product-compliance
Lightning Source LLC
Chambersburg PA
CBHW071700210326
41597CB00017B/2258
* 9 7 8 1 8 8 7 2 0 1 9 0 2 *